1993

Fractals for the Classroom:
Strategic Activities Volume Two

Heinz-Otto Peitgen
Hartmut Jürgens
Dietmar Saupe

Evan Maletsky
Terry Perciante
Lee Yunker

Fractals for the Classroom:
Strategic Activities Volume Two

National Council of
Teachers of Mathematics

Springer-Verlag

Heinz-Otto Peitgen
Institut für Dynamische Systeme
Universität Bremen
D-2800 Bremen 33
Federal Republic of Germany and
Department of Mathematics
Florida Atlantic University
Boca Raton, FL 33432
USA

Hartmut Jürgens
Institut für Dynamische Systeme
Universität Bremen
D-2800 Bremen 33
Federal Republic of Germany

Dietmar Saupe
Institut für Dynamische Systeme
Universität Bremen
D-2800 Bremen 33
Federal Republic of Germany

Evan Maletsky
Department of Mathematics and
    Computer Science
Montclair State College
Upper Montclair, NJ 07043
USA

Terry Perciante
Department of Mathematics
Wheaton College
Wheaton, IL 60187-5593
USA

Lee Yunker
Department of Mathematics
West Chicago Community High School
West Chicago, IL 60185
USA

TI-81 Graphics Calculator is a product of Texas Instruments Inc.

Casio™ is a registered trademark of Casio Computer Co. Ltd.

Printed on acid-free paper.

Negatives supplied by the authors.
Printed and bound by John D. Lucas Printing Co., Baltimore, MD.
Printed in the United States of America.

9 8 7 6 5 4 3 2 1

ISBN 0-387-97554-3 Springer-Verlag New York Berlin Heidelberg
ISBN 3-540-97554-3 Springer-Verlag Berlin Heidelberg New York

Preface

The same factors that motivated the writing of our first volume of strategic activities on fractals continued to encourage the assembly of additional activities for this second volume. Fractals provide a setting wherein students can enjoy hands-on experiences that involve important mathematical content connected to a wide range of physical and social phenomena. The striking graphic images, unexpected geometric properties, and fascinating numerical processes offer unparalleled opportunity for enthusiastic student inquiry.

Students sense the vigor present in the growing and highly integrative discipline of fractal geometry as they are introduced to mathematical developments that have occurred during the last half of the twentieth century. Few branches of mathematics and computer science offer such a contemporary portrayal of the wonderment available in careful analysis, in the amazing dialogue between numeric and geometric processes, and in the energetic interaction between mathematics and other disciplines. Fractals continue to supply an uncommon setting for animated teaching and learning activities that focus upon fundamental mathematical concepts, connections, problem-solving techniques, and many other major topics of elementary and advanced mathematics.

It remains our hope that, through this second volume of strategic activities, readers will find their enjoyment of mathematics heightened and their appreciation for the dynamics of the world increased. We want experiences with fractals to enliven curiosity and to stretch the imagination.

A third volume already in progress emphasizes iterated function systems. Like the first two, it is driven by the desire to relate new and significant mathematical developments to the teaching-learning enterprise. The international team writing these volumes of strategic activities in fact exemplifies the wonderful results that can come from collaboration between researchers and teachers. We hope that the synergistic and very positive effect that we have found in our collaboration with each other will in some way be conveyed through our writing and ultimately reflect itself in similar outcomes between teachers and their students.

We are very grateful that Mitchell Feigenbaum accepted our invitation to write the foreword addressing his discovery of the period-doubling route from order to chaos and its universality in dynamical systems. As he explains, these fascinating phenomena were originally conceived using merely a simple programmable pocket calculator. Today such devices are standard equipment in most mathematics classrooms and, in this volume, we retrace some of the intriguing explorations that led the scientific world to this new understanding.

This volume of *Strategic Activities* for *Fractals for the Classroom* has been prepared by the authors using the TEX and LATEX typesetting systems. We owe our gratitude to our students Torsten Cordes and Lutz Voigt who have produced the immense number of figures with great expertise and patience.

Evan M. Maletsky
Terry Perciante
Lee E. Yunker
USA, 1992

Heinz-Otto Peitgen
Dietmar Saupe
Hartmut Jürgens
Germany, 1992

Authors

Hartmut Jürgens. ∗1955 in Bremen (Germany). Dr. rer. nat. 1983 at the University of Bremen. Research in dynamical systems, mathematical computer graphics and experimental mathematics. Employment in the computer industry 1984–85, since 1985 Director of the Dynamical Systems Graphics Laboratory at the University of Bremen. Author and editor of several publications on chaos and fractals.

Evan M. Maletsky. ∗1932 in Pompton Lakes, New Jersey (USA). Ph. D. in Mathematics Education, New York University, 1961. Professor of Mathematics, Montclair State College, Upper Montclair, New Jersey, since 1957. Author, editor, and lecturer on mathematics curriculum and materials for the elementary and secondary school, with special interest in geometry. First Sokol Faculty Award, Montclair State and 1991 Outstanding Mathematics Educator Award, AMTNJ.

Heinz-Otto Peitgen. ∗1945 in Bruch (Germany). Dr. rer. nat. 1973, Habilitation 1976, both from the University of Bonn. Research on nonlinear analysis and dynamical systems. 1977 Professor of Mathematics at the University of Bremen and from 1985 to 1991 also Professor of Mathematics at the University of California at Santa Cruz. Since 1991 also Professor of Mathematics at the Florida Atlantic University in Boca Raton. Visiting Professor in Belgium, Italy, Mexico and USA. Author and editor of several publications on chaos and fractals.

Terence H. Perciante. ∗1945 in Vancouver (Canada). Ed. D. in Mathematics Education, State University of New York at Buffalo, 1972. Professor of Mathematics at Wheaton College, Illinois, since 1972. Received a 1989–90 award for Teaching Excellence and Campus Leadership from the Foundation for Independent Higher Education. Author of several books, study guides and articles at the college level.

Dietmar Saupe. ∗1954 in Bremen (Germany). Dr. rer. nat. 1982 at the University of Bremen. Visiting Assistant Professor of Mathematics at the University of California at Santa Cruz, 1985–87 and since 1987 Assistant Professor at the University of Bremen. Research in dynamical systems, mathematical computer graphics and experimental mathematics. Author and editor of several publications on chaos and fractals.

Lee E. Yunker. ∗1941 in Mokena, Illinois (USA). 1963 B. S. in Mathematics, Elmhurst College, Elmhurst, Illinois. 1967 M. Ed. in Mathematics, University of Illinois, Urbana. 1963–present: Mathematics Teacher and Department Chairman, West Chicago Community High School, West Chicago, Illinois. Member of the Board of Directors of the National Council of Teachers of Mathematics. Presidential Award for Excellence in Mathematics Teaching, 1985. Author of several textbooks at the secondary level on geometry and advanced mathematics.

Table of Contents

Unit 6   **THE MANDELBROT SET**

Connections to the Curriculum

Students can best see the power and beauty of mathematics when they view it as an integrated whole. These charts show the many connections these strategic activities have to established topics in the contemporary mathematics program.

The National Council of Teachers of Mathematics, in its *Curriculum and Evaluation Standards for School Mathematics*, stresses the importance of mathematical connections:

> "The mathematics curriculum should include investigation of the connections and interplay among various mathematical topics and their applications so that all students can
>
> - recognize equivalent representations of the same concept;
> - relate procedures in one representation to procedures in an equivalent representation;
> - use and value the connections among mathematical topics;
> - use and value the connections between mathematics and other disciplines."

Unit 4

Connections	4.1	4.2	4.3	4.4	4.5	4.6	4.7	4.8	4.9	4.10	4.11	4.12	4.13
Linear Functions	●	●	●	●	●	●							
Quadratic Functions		●	●		●		●	●	●	●	●	●	●
Function Composition		●	●		●		●	●	●	●	●	●	●
Evaluating Functions		●		●	●	●	●	●	●	●	●	●	●
Transformations			●	●	●								●
Mappings			●		●		●	●		●	●		
Numerical Patterns	●	●		●		●	●	●	●	●	●	●	●
Geometric Patterns	●		●	●		●	●	●	●	●	●	●	●
Visualization	●		●	●	●	●	●	●	●	●	●	●	
Sequences and Series		●											
Convergence	●	●				●	●		●	●	●	●	
Limit Concept		●					●		●	●	●	●	
Absolute Value				●		●							
Graphing	●		●	●	●	●	●	●	●	●		●	●
Slope	●			●			●		●	●		●	
Graphing Calculator								●	●		●	●	●

Unit 5

Connections	5.1	5.2	5.3	5.4	5.5	5.6	5.7	5.8	5.9	5.10	5.11	5.12	5.13
Linear Functions		●											
Quadratic Functions								●	●	●	●	●	●
Piecewise Functions		●		●	●	●	●				●		
Function Composition		●		●	●	●	●	●	●	●	●	●	●
Evaluating Functions		●		●	●	●	●	●	●	●	●	●	●
Transformations	●	●		●	●	●	●		●	●	●		
Mappings	●	●		●	●	●	●		●	●	●	●	●
Numerical Patterns	●	●	●	●	●		●	●	●	●	●	●	●
Geometric Patterns	●	●		●	●	●	●	●	●	●	●	●	●
Visualization	●	●		●	●	●	●	●	●	●	●	●	●
Sequences and Series			●										
Convergence								●				●	●
Limit Concept								●				●	●
Binary Numbers			●	●	●	●	●						
Rational Numbers			●		●								
Irrational Numbers					●								
Coordinate Geometry								●	●	●	●	●	●
Graphing					●	●	●	●	●	●	●	●	●
Graphing Calculator								●	●	●	●	●	●

Unit 6

Connections	6.1	6.2	6.3	6.4	6.5	6.6	6.7	6.8	6.9	6.10	6.11	6.12	6.13
Quadratic Functions				●	●	●	●	●		●	●	●	●
Piecewise Functions		●	●		●								
Function Composition		●	●	●	●	●				●	●	●	●
Evaluating Functions		●		●	●	●	●	●		●		●	●
Transformations							●						
Mappings							●						
Numerical Patterns			●										
Geometric Patterns	●		●	●						●	●	●	●
Visualization	●		●	●	●	●	●	●			●	●	●
Limit Concept			●										
Conic Sections	●												
Complex Numbers									●	●	●	●	●
Coordinate Geometry	●	●	●				●		●	●	●	●	●
Graphing				●	●	●	●	●	●	●		●	●
Graphing Calculator						●		●		●		●	

Foreword

The outcomes of dynamical action — chaotic or near-random evolution on and of fractal objects — are phenomena that are only recently within the range of human understanding. This is so neither because of human minds nor computers alone. Rather, the understandings we have attained have arisen from a fine collaboration of the two. It is insufficient to *think* when even the elementary consequences of those thoughts are beyond human power to explore for guidance, and it is pointless to *compute* where neither intuition nor thought inheres. Together, wonderful new understandings of age-old observations have come to light. Moreover, the "understandings" don't merely stay in the computer. Once perceived there, they have been formulated as humanly manipulable ideas, so that, by and large, the computer can be forgotten, with the newly informed humans swiftly beating the brute force of the computer to the finish. Often these ideas can be altogether elementary: one simply hadn't known how to obtain them — nor even had sensed their significant presence — without numerical support. That there is new knowledge of an elementary and significant sort makes the study of dynamics altogether exciting and pleasurable, and appropriate for the early introduction

Mitchell J. Feigenbaum

into education. Thus, there is very much to learn analytically by simple manipulations, and much to learn computationally with a resource no grander than a pocket calculator.

The use of computers in this study is one of exploration, to discover which intuitions and ideas are at all insightful, and so then sharpen them to discover which directions of development are apt to be fruitful for further inquiry. As the exploration grows deeper — happily sometimes almost immediately — novel, unsuspected things can indeed transpire, and entirely new paths of exploration can emerge. Let me describe this process from my own experience.

Although no proof had been given, it was known by the early 1970's that some simple systems possessed a cascade of thresholds for increasingly erratic behavior. Prototypically, consider a watch with a +/− speed adjustment screw, with the following properties holding. As one turns the screw towards +, at some point, the watch starts beating out a constant rhythm of first a short tick and then a long one. After some further turning of the screw, the rhythm suddenly changes to a repetition of $4 = 2^2$ different length ticks. And with successively smaller amounts of further turning, 8, 16, 32, 64, ... tick patterns appear, until finally, before having reached the full + position, the watch ticks with the length of any one tick different from that of any other.

I worked on this phenomenon of period doubling for a year, starting in the summer of 1975. I first constructed a theoretical picture that made the cascade sensible, but failed to understand how things worked by the end of the cascade. Returning home, I attempted to explore my theory numerically, using the only computer I had used, an HP-65 programmable calculator. Failing to gain better insight, I decided to see what happened if I used the results from a numerical solution for repetition periods of length that were powers of two. This meant that I first had to find the screw settings for which such behaviors were to be observed. The clock was modelled by a particularly simple (quadratic) expression that determined the length of the next tick in terms of the present one, with the angle of the screw represented by a constant, say "r". The $n^{\text{th}}r$ value is that for which the period of ticks is the nth power of two. For $n > 2$ the values can only be obtained numerically. For $n = 6$ the calculator took about a minute, while it grew increasingly difficult to start it off right so as not to have gotten a spurious result. Thus, after I wrote down each value, I had time to stare at the numbers. It became immediately clear (by estimating the ratios of successive differences) that they were geometrically converging. Here the HP had proven infinitely superior to any real computer: several people had obtained these numbers, and hadn't noticed the fact, almost certainly because they had never looked at what their computers produced! Interactive slowness can be a cardinal virtue. What the convergence observation meant was that just when the computer's ability was being compromised, it became increasingly easy for a human to get the result.

Where did a geometric convergence come from? I didn't know, nor did the results prove helpful in my theoretical investigation. The problem lay fallow for two months until I realized that not only the simple quadratic formula for new ticks, but trigonometric ones as well were supposed to have an infinite cascade of doublings. My theory, however was only natural for quadratic ones, so that I reasoned the trigonometric one didn't double or my ideas were wrong. Resorting to the HP showed indeed that it doubled (and the theory too was right, but not very insightful.) Now, $n = 6$ took over five minutes. But, lo and behold, the numbers again geometrically converged. And much, much more wonderful: They converged at *exactly the same rate* as did those for the quadratic! So, with limited, but some understanding of what I was exploring, a serendipitous jewel shone forth — because it meant that all these dynamical problems behaved in a *universal* way, totally independent of the equations of their motion.

It took another two months to realize that the story of this number was not in the screw, but in the ticks themselves. These too, in their increasing diversification, show off geometric convergences, although with yet another new universal constant. Moreover, I discovered that the setting for all these thoughts was in "functional composition operators". Now all that was necessary was to find the appropriate mathematics and its practitioners. There was one problem: The practitioners knew that the mathematics didn't exist. So, it now became necessary to learn how to understand and solve such infinite dimensional problems numerically. The HP was at last beyond its depth, and it was time to use a seriously powerful computer. (With a good problem to work on, and if you don't waste your time on fancy input-output, it is very easy to learn in a day's time. Throw out any tutorial manuals that make you first do input-output and sort embarrassing trash!). The process of doing this was absolutely exhilarating, entailing having an idea, writing a hundred line program to explore it, learning from that how to improve the idea, and back and forth in rapid-fire succession, producing five to ten odd programs per day. (I was immensely excited: these were twenty odd working hour days, and a slower machine would have been frustrating in the extreme.)

The theory was finished by the end of summer, 1976. The entire picture was all spelled out, but any rigorous proof missing. Indeed, apart from a computational proof elaborating on my methods, no human one was forthcoming until two years ago. The proof is arduous in the extreme, while still not covering all of the picture. Clearly such a proof would never have been undertaken had one not known that there was something to prove, and what was to be proven. (Sullivan constructed his proof over an almost ten year period.) Without numerics, I firmly believe the knowledge of these phenomena would still be absent.

Dynamics is rich soil in which decidedly new phenomena remain to be found and understood. From it, the fractals on which the motion occurs have not one or two different scales, but always infinitely many. Beyond one dimension almost nothing is understood of their properties if these fractals arise in natural physical contexts. All the mathematics is awaiting to be discovered. Indeed, one can solve so few realistic hard problems because the approximation methods that exist are mostly inappropriate, while brute-force computation is well beyond any machine that yet exists. Is the solar system stable? Almost certainly not, but this is a numerically informed surmise only. Theory is totally absent. No methods exist to contemplate realistically, however simplified, what kinds of behaviors appear in a vastly interconnected dynamical system, such as a brain. Two decades ago, one couldn't have honestly approached any such problem. Now at least hints exist, with the ideas behind them often of an elementary character. Some of them are well within your grasp, as you're about to discover.

Mitchell J. Feigenbaum
Toyota Professor

The Rockefeller University
March 1992

Unit 4
Iteration

KEY OBJECTIVES, NOTIONS, and CONNECTIONS

The activities in this unit transform the traditional mathematical content of the typical secondary curriculum into a new, dynamic, visual, geometric world. Here we see things constantly change as we search for patterns in iterative behaviors. Graphical iteration produces paths that staircase or spiral in to specific points that serve as attractors and out from other points that play the role of repellers. Some intervals compress through graphical iteration so that they shrink even the smallest of errors, while others expand through iteration, causing similar small errors to explode into large ones. Underlying these characteristics is the key question: When is the iteration behavior predictable and when is it unpredictable?

The process of iteration is developed from an intuitive perspective that connects it to the familiar notion of composition of functions. Attention is focused on the discrete aspects of both graphical and numerical iteration as a recursive process. We follow the changes an initial point undergoes as it is iterated through the same function over and over again. Often a great deal can be learned about how a dynamical system operates by studying the behavior of only a few points. The experience can be both exciting and enriching. It will challenge the intuition when the discovery is made that similar quadratic functions can exhibit behaviors that are dramatically different.

Connections to the Curriculum

The material covered in these strategic activities form an integral part of a contemporary mathematics program. They may be used as a single unit on this topic or integrated into the existing curriculum through those areas to which they are connected.

PRIMARY CONNECTIONS:

Quadratic Functions	Slope
Geometric Patterns	Visualization
Numerical Patterns	Limit Concept
Function Composition	Convergence
Mappings	Graphing
Evaluating Functions	

SECONDARY CONNECTIONS:

Series and Sequences	Absolute Value
Linear Functions	Graphing Calculator
Transformations	

Underlying Notions

Composition of Functions The formation of a composite function $f(g(a))$ is one function g operating on an argument a followed by a second function f operating on a new argument $g(a)$.

Iteration Iteration is the process of repeatedly forming the composition of a function with itself.

Graphical Iteration Graphical iteration is the process of visually representing the iteration of a function.

MATHEMATICAL BACKGROUND

The Bigger Picture

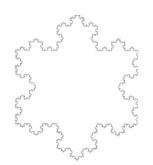

von Koch Snowflake

Mandelbrot Set

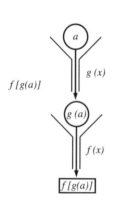

$f[g(a)]$

a

$g(x)$

$g(a)$

$f(x)$

$f[g(a)]$

The process of iteration is so central to any study of fractals that it is difficult to overstate its importance. For example, in order to create the von Koch snowflake described in Unit 3 of the first volume, an iterative replacement process is used in which one segment is replaced by four segments, each 1/3 the length of the original. In the Mandelbrot set and its associated Julia sets, described in Unit 6 of this second volume, iteration is the fundamental factor used in determining whether a point belongs to one of these sets or not. Iteration also plays a major role in determining the coloring of the points surrounding these sets.

The iteration process in this unit begins with linear functions but focuses primarily on iterating the family of quadratic functions of the form $f(x) = ax(1 - x)$. This general quadratic function, with real values x, is mathematically equivalent to the quadratic function $z^2 + c$ used, with complex values z, to determine the Mandelbrot set. This unit will concentrate on iterating real numbers in the closed interval from 0 through 1. Later, iteration of the real numbers will be expanded beyond this closed interval. Eventually, iteration of numbers in the complex plane will be studied in relation to the points in the Mandelbrot and Julia sets.

Finally, it must be noted that, without iteration, there would be no dynamic interplay between points and functions, there would be no Mandelbrot and Julia sets, and there would be no chaos. The power of iteration goes far beyond fractals. It is an important, underlying mathematical tool for modeling natural phenomena found in our real world, such as weather change, population growth, chemical reactions, fluid flow, and heart beats.

Iteration as a Composition

An intuitive approach to the concept of composition can be achieved through the notion of a function machine. The function machine accepts a raw material or input (domain) and produces a product or output (range). The next machine accepts an input or raw material (output from the first machine) and produces a new output or product.

Consider an analogy in the canning of orange juice, where the first machine accepts oranges (input) and produces orange juice (output) by squeezing the oranges. The second machine accepts the orange juice (input) and cans it, producing canned orange juice (output). This approach helps those unfamiliar with the concept of a function to internalize the process. An interesting question to ask is this: What happens if the order of the machines in the orange juice canning process were accidentally reversed?

Recursive Compositions

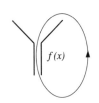

Iteration can be introduced as a recursive process. It can be viewed as the composition of a function with itself, where the output at one step in the process becomes the input at the next step of the very same process, over and over again.

Iteration is often referred to as a feedback loop. A simple example of this can be illustrated on a calculator. Take the square root of some positive number with your calculator and feed the output back into the calculator, taking the square root of the new result. Then repeat the entire process over and over again.

Arithmetic and geometric sequences are generated by iterative processes. The next term in each sequence is generated by either repeatedly adding a constant to the previous term or repeatedly multiplying it by a constant.

Iteration as a Graphical Process

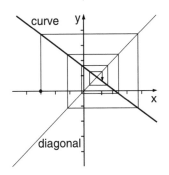

Iteration is presented as a simple algorithmic process of drawing nothing more than vertical and horizontal segments, first to the graph of a function and then to the diagonal $y = x$, which reflects it back to the graph again. Repeating the process over and over generates a continuous path of alternating vertical and horizontal segments. Much can be learned about the behavior of the function from the visual shape of the iteration path.

Fixed points of a function can be identified by intersections of the graph of the function with the diagonal. Their nature can be attracting, repelling, or indifferent. Examining graphical iteration in the neighborhood of a fixed point is a visual tool to determine the respective property. Some paths staircase or spiral in, indicating attracting fixed points, while others staircase or spiral out, indicating repelling fixed points. These behaviors are closely tied to the values of the slope of the function at the fixed points.

Slope		
	$m < -1$	Repel and spiral out
	$-1 < m < 0$	Attract and spiral in
	$0 < m < 1$	Attract and staircase in
	$1 < m$	Repel and staircase out

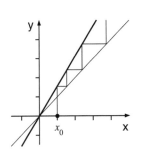

The diagonal $y = x$ acts as the recursive mechanism, reflecting the output or range value from one step of the process back into the system as the input or domain for the next step in the process. This reflection transformation property of graphical iteration is simple and yet powerful, taking full advantage of our visual senses.

Expansion and Compression

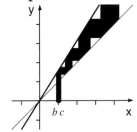

Quadratic Function

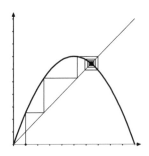

Attractors

Periodic Points

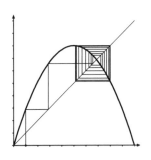

Predictability

Another important behavior deals with the effect iteration has on a small subinterval, which models error in the input. With interval expansion, small errors expand, through iteration, into large ones. With interval compression, small errors contract and converge to 0.

Slope $|m| > 1$ Interval expansion
$0 \le |m| < 1$ Interval compression

Numerous opportunities are provided for practicing the process of graphical iteration on both linear and quadratic functions. This hands-on experience with pencil and ruler is essential if one is to internalize the process.

One family of quadratic functions that receives special attention has the form $f(x) = ax(1 - x)$. These functions exhibit, under iteration, behaviors central to the understanding of chaos in a dynamical system. While substantial attention is given here to iterating these quadratic functions where $a = 2.8$, 3.2, and 4.0, it is important to emphasize that these parameter values are only representative of the many functions with $1 \le a \le 4$ that exhibit similar behaviors.

Quadratic functions in the form $f(x) = ax(1 - x)$ are mathematically equivalent to those in the form $f(x) = x^2 + c$. Their iteration behaviors are essentially the same. In Unit 6, functions in the form $f(z) = z^2 + c$, with z complex, lead directly to the Mandelbrot and Julia sets.

Iterating the function $f(x) = 2.8x(1 - x)$ for most values in the closed interval [0,1] produces an interesting result. It has the long term iterative behavior of always leading to the same single value, 0.6429... This convergence on a single value, called an attractor, occurs for those functions with a parameter value of a anywhere between 1 and 3.

Iterating the function $f(x) = 3.2x(1 - x)$ for most values in the interval [0,1] produces a different result. It has the long term iterative behavior of always leading to the same two different values, 0.799455... and 0.513044... This periodic, oscillating long term behavior, called a period-two attractor, will occur for those functions with parameter values of a between 3 and 3.4494...

Iterating one of the values of a period-two attractor will always produce the other. Often the period of an attractor is greater than two, meaning that the iteration cycle takes more than two iterations before it returns to the same periodic point for a second time. Many other pre-periodic points eventually iterate into points that are periodic.

The two functions $f(x) = 2.8x(1 - x)$ and $f(x) = 3.2x(1 - x)$ have very predictable iterative behaviors and illustrate stable dynamical systems. Their long term iterative behaviors are well known and always lead to the same attractors, irrespective of the initial values.

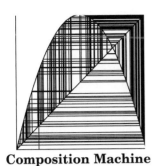

Composition Machine

However, another similar function, $f(x) = 4x(1-x)$, produces quite different results for initial values in the interval [0,1]. Here the long term iterative behavior is erratic and unpredictable, with no recognizable pattern. Two points, very close to each other, generate radically different iterative sequences. This phenomenon, known as sensitive dependence on initial conditions, means that small initial errors will expand into major ones, making the results totally unreliable. So pronounced is this chaotic behavior that you can experience it when repeating graphical iteration for exactly the same initial values.

This unit ends with a geometric process for finding the composition $g(f(x))$ of the two functions $f(x)$ and $g(x)$. While such a composition can always be found through algebraic substitution, the visual impact of this composition machine is very powerful. Points on one function $f(x)$ are reflected off the diagonal $y = x$ and through the second function $g(x)$.

It turns out that fixed points (resp. periodic points of period two) in the iteration of $f(x)$, appear as intersection points between the graph of $f(x)$ (resp. the graph of $f(f(x))$) and the diagonal, $y = x$. The periodic points cannot be found precisely using graphical iteration, even though the behavior indicates their existence. However, periodic points of period two can be read directly from $f(f(x))$.

Iteration using Technology

The graphing calculator is a very powerful tool for visualization and exploration. Program code is presented in this unit for both the Texas Instruments and Casio graphing calculators. Attention is given to comparing the iterative behaviors viewed on the calculator to the same iterations done with pencil and ruler. The calculator is also used to do rapid numerical iterations on functions so behaviors can be explored through the study of numerical patterns.

Additional Readings

Chapters 10 and 11 of *Fractals for the Classroom, Part Two*, H.-O. Peitgen, H. Jürgens, D. Saupe, Springer-Verlag, New York, 1992.

USING THE ACTIVITY SHEETS

4.1 Spiral and Staircase

Specific Directions. This activity introduces the basic idea of graphical iteration through an informal algorithmic process of repeatedly moving back and forth between a curve and the diagonal with connected vertical and horizontal segments. At this point, the process produces only staircases and spirals, each leading in to a fixed point as an attractor or away from a fixed point as a repeller. Linear functions are used for their consistent iteration behavior. Details of the graphical iteration procedure and its justification come later.

Implicit Discoveries. The slope at the fixed point determines the iteration behavior about that point. Since all functions used here are linear, the behavior is not dependent on the location of the initial point. The one exception is when the initial point is the fixed point itself.

4.2 Composition of Functions

Specific Directions. The concept of a function is difficult for some to grasp. The notion can be developed from an intuitive perspective as a rule or function applied to each input or first coordinate x to produce a corresponding output or second coordinate $f(x)$. The concept of a composition of functions is equally difficult for some to understand. This is often a notational problem. To execute $f(g(x))$, use g first and then f. Start with a domain value a for x and use it to obtain the corresponding range value, $g(a)$. Then use $g(a)$ as the next domain value and use it to obtain the corresponding range value, $f(g(a))$. The graphics of a funnel helps communicate the idea of this input-output relationship between x and $f(x)$. When combined, they illustrate the composition of functions in such a way as to make the ordering of the functions visible.

The iterative process arises when repeatedly forming the composition of a function with itself. Many number patterns such as arithmetic or geometric sequences can be described and generated by such an iterative process. It is important to see and experience these familiar topics through the eyes of a new concept such as iteration and to connect them to specific iteration behaviors.

Implicit Discoveries. Several questions draw attention to the fact that the composition of functions is not always commutative. Reversing the order of the functions will often cause different results to occur. The function $f(x) = \sqrt{x}$ used here gives the numerical counterpart to the geometric iteration behavior of the staircase around the attractor, 1. The function $f(x) = x^2$ used here gives the numerical counterpart to the geometric iteration behavior of the staircase around the repeller, 1.

4.3 Graphical Composition of Functions

Specific Directions. The study of iteration and chaos are built on the concept of the composition of functions. Activity 4.2 presented composition from an algebraic point of view. This activity builds on the idea from a geometric vantage point. Piecewise and non-linear, as well as linear functions, are used. Be sure to establish the function transformation as a physical process of moving up from x to the graph of the function (evaluation), left or right to the diagonal (transfer), and then out as $f(x)$ (reflect). Stacked together, one on top of another, a visual picture of the composition of functions appears. In evaluating $g(f(x))$, the first output $f(x)$ becomes the input for $g(x)$. The composition process can be traced through from function to function. The final part of the activity introduces this process of composition with a single repeating function and then shows how the step-by-step process can be combined onto a single graph. The repeated two-step geometric iteration process is vertical (evaluate) and horizontal (feedback). This generates the successive iterates along the x-axis. It is the mechanical algorithm that will become the cornerstone of the graphical iteration used throughout the remainder of the unit.

Implicit Discoveries. Stress the importance of the connection between the geometric process of graphical iteration and the algebraic process of function composition. Note the importance of the diagonal $y = x$ as the reflection transformation mechanism that gives the feedback, mapping every $f(x)$ value back as the next iterate value.

4.4 Fixed Points as Attractors and Repellers

Specific Directions. This activity establishes more fully the concept of fixed points as repellers first introduced in Activity 4.1. Several steps of graphical iteration are performed and the behavior observed as it relates to the region around the fixed point. Since the fixed point must occur where the function meets the diagonal, its x- and y-coordinates are always the same. Critical to the iteration behavior about that point is the slope of the function at that point. Use the completed drawings of staircase and spiral behavior to discuss convergence to a fixed point serving as an attractor or divergence away from one that serves as a repeller. Stress the visual impact of the iteration graphs and seek verbal descriptions of comparison and contrast.

Implicit Discoveries. With these linear functions, it should be clear how the slope determines the behavior. Repelling occurs with slopes less than -1 and greater than 1. Attracting occurs with slopes between -1 and 1. Spiral behavior occurs with negative slopes and staircase behavior occurs with positive slope. Note the behavior at the points where the slope is -1, 0, and 1.

4.5 Solving Equations and Finding Preimages

Specific Directions. Evaluating a function $y = f(x)$, given some value for x, consists of a direct substitution operation. The algebraic form facilitates direct computation of y from x. On the other hand, solving an equation requires reversing the process to find x from y and this procedure can be more complicated. A similar situation occurs in the composition of functions. This activity views that process both algebraically and geometrically.

To work back from $f(g(x))$ to x, first solve $z = f(y)$ for y and then solve $y = g(x)$ for x. We say x is the preimage of $y = g(x)$, and y is the preimage of $z = f(y)$. The degree of the function affects the number of preimages to be found at that level.

When viewing functions geometrically, the preimage or preimages x are found from y by moving horizontally to the curve from the y-axis and then vertically to the x-axis. Of course, as with algebraic solutions, more than one preimage will occur if the horizontal line meets the curve at more than one point. To undo the composition of functions geometrically, work back from the top of the stack, being careful to take note of any places where the paths may split.

Implicit Discoveries. One of the problem-solving strategies that receives a great deal of attention in these times is that of working backwards. The discussion of preimages in this activity will help to develop that skill.

4.6 Intervals and Errors

Specific Directions. Accurate drawings are essential in studying error behavior. We use small subintervals to model error in the input values. The iterated interval, while still in the form of a staircase or spiral, also exhibits the behavior of expansion or compression. Attractors are tied to compression and repellers to expansion. When the small interval is viewed as an error, one quickly sees how slope can, in cases where it is less than -1 or greater than 1, cause a small error to grow into a large one. The concern to note is not so much the size of the error as it is the effect the error has on reliability of the values generated.

Implicit Discoveries. It is the slope of the function at a given point that determines the nature of the iteration error at that point. Very large positive or negative slopes create rapid error expansion while errors are dramatically reduced for slopes near 0.

It is important to note that absolute errors can be large while the related relative errors are small. However, a large relative error means a substantial portion of the reported figure may be suspect. This means that the analysis of natural phenomena through iteration can be highly misleading producing unpredictable results in cases where interval expansion occurs.

4.7 Iterating $f(x) = ax(1 - x)$: **Attractors**

Specific Directions. The graphical iteration path starts on the x-axis with the initial value x_0. From there on, it moves back and forth between the function and the diagonal. To find the successive iterates, extend successive vertical segments in the iteration path down to the x-axis. The iterates are read off as the values of these intersection points on the x-axis. Follow the iteration path carefully, especially with the spiraling behavior, to be certain that successive iterates are listed in the correct order. In this activity, the intersection point of the parabola and the diagonal serves as attractor or repeller, depending on the value of the parameter a in the quadratic function $f(x) = ax(1 - x)$. Focus on the changing behavior of the iteration path as the parameter a changes.

Implicit Discoveries. Some iteration patterns will spiral in to an attractor faster than others, depending on the choice of initial values. The concept of an attractor is the result of the long term iterative behavior of a function and may not always appear as an immediate result of the iterative process. When the parameter a in $f(x) = ax(1 - x)$ is 2.8, the single attractor of period one (fixed point) is 0.6428... This is the x-value of the fixed intersection point of the function and the diagonal.

When the parameter a is 3.2, there is a cycle of period two: 0.5130... $\rightarrow$ 0.7994... $\rightarrow$ 0.5130... $\rightarrow$ 0.7994... $\rightarrow$ $\cdots$ Iteration paths typically spiral towards a square box defined by these values. The fixed point of the function now serves as a repeller. The attractor now consists of the two points identified above.

4.8 Iterating $f(x) = ax(1 - x)$: **Chaos**

Specific Directions. This activity requires the actual drawing of the iteration path from several different initial points for $f(x) = 4x(1 - x)$. Again, extend successive vertical segments down to intersect the x-axis and read off the values of the iterates. Comparing results for the first activity sheet with others will likely produce strikingly different iteration patterns. Surprisingly, this is the expected outcome! It is our first experience with chaos in an iterated dynamical system. In a classroom experiment, it is interesting to compare the results obtained by students. For example a table of the various outcomes of the last computed value x_7 will reveal the surprise. The results will likely be distributed throughout the entire unit interval. This is a consequence of the *sensitivity* to small errors in the iteration of the quadratic function $4x(1 - x)$, an important topic addressed in Activity 4.11.

Implicit Discoveries. We discover that no matter how carefully we draw our iteration patterns, getting them perfect is impossible. Moreover, these imperfections are magnified in the iteration process so that at the end, after some iterations, the errors can completely overshadow the true results. This is a rather unsettling feeling, given our past results with other parabolas. We learn from this activity that even closely related functions can behave, under iteration, in dramatically different ways. The more important question is whether or not these behaviors can be predicted before hand? Unit 5 addresses this question in detail.

4.9 Graphical Iteration through Technology

Specific Directions. The focus of this activity is to use technology to generate the graphical iteration so that full attention can be given to the iteration behavior. Take care in entering the program. Use question 1 with the program and verify the behaviors already shown earlier. Question 2 leads to the identification of short-term iteration behaviors and how they change as the value of the parameter a increases from 1 to 4. In these two cases, the program shows the iterations from the initial iterate. The changes made in question 3 activate a delayed procedure of 100 iterations before plotting begins. This allows a close examination of the long term dynamical behavior. Question 4 shows how dramatically the behavior can change as the parameter a approaches 4. One can see the transition from order to chaos.

Implicit Discoveries. We quickly discover how powerful the graphing calculator is in exploring graphical iteration. However, the scene of the transition from order to chaos is not a smooth one. There is a period-doubling sequence from 1 to 2 to 4 to 8 and so on. But as the parameter a approaches 4, attractors with diverse periods appear to alternate back and forth with chaotic behavior in a more complex fashion. More details on this issue appear in Activity 5.13.

4.10 Expansion and Compression in $f(x) = 4x(1-x)$

Specific Directions. The iteration behavior for $f(x) = 4x(1-x)$ is erratic, chaotic, and unpredictable. Linear functions have stable, predictable behaviors, as do many of the parabolas studied earlier. The key to the difference lies in the nature of the slope of this function over the interval [0,1]. As the parameter a gets close to 4, a larger and larger portion of the interval has a slope less than -1 or greater than 1. With that introduction, ask these questions:

What effect does this have in relation to subinterval expansion?
What effect does iteration have on small errors if there is substantial interval expansion?
What effect do major deviations from slopes of 1 and -1 have on the rate of expansion?
What effect do the answers to the above questions have on predicting iteration behavior?

Implicit Discoveries. The answers to the questions listed above reveal the cause of the chaotic behavior found when iterating $f(x) = 4x(1-x)$. Much of the unit interval [0,1] is in interval expansion because the slope is greater than 1 or less than -1. Since much of the curve is steep, the rate of expansion is great. All these factors lead to a state where very small errors rapidly magnify and explode, taking control and creating the chaotic behavior that makes any kind of prediction impossible.

4.11 Sensitivity

Specific Directions. The graphing calculator is the tool used here to explore sensitivity in greater detail. Questions 1–7 compare numerical iterates and their corresponding behaviors through the first seven stages for different initial values, different parameter values, and different computing precision. The results show how small rounding errors alter results most dramatically under $f(x) = ax(1-x)$. In the first run of question 9, the "exact" values using 10 place precision are computed and compared to the second run with 3 places of accuracy. This demonstrates that the two iteration paths clearly diverge from each other. The same phenomenon occurs in question 11 when considering a very small difference in the initial value x_0 while using the same high precision of 10 digits.

Implicit Discoveries. Through this activity, we see just how sensitive and dependent the function $f(x) = ax(1-x)$ is, under iteration, to very small changes in initial or subsequent conditions. This problem, so vividly illustrated on the graphing calculator is, in fact, inherent in all technology. At best, every computer must deal with the same constraints of finite digit arithmetic. The implications that follow from this observation are profound. Many systems, such as those used to predict the weather, are susceptible to this same kind of erroneous information.

4.12 Iterating Quadratics

Specific Directions. Thus far, we have considered quadratic functions of the form $f(x) = ax(1-x)$ over the interval [0,1] with $1 \leq a \leq 4$. The first questions in this activity contrast the iteration behavior for two parabolas of the same form for $a = 0.5$ and $a = 5$, this time for all real values for x. Be sure to recognize that a new behavior arises when a point within the interval [0,1] escapes to (negative) infinity.

Look for the new behavior that arises on the second page when parabolas of the form $f(x) = x^2 + c$ are used. Questions 9 and 10 illustrate a new behavior that first appears as a staircase in toward some attractor but then abruptly changes and begins to staircase out to infinity. Question 11–13 investigates this property on the graphing calculator with a new program for iterating these functions.

Implicit Discoveries. Certain points within the interval [0,1] can escape to (negative) infinity when iterated under the function $f(x) = ax(1 - x)$, but only when $a > 4$. These points occur when an iterate reaches the subinterval in x that produces an $f(x)$ that rises above the value 1. The behavior shown in questions 9 and 10 occurs , for example, when the function $f(x) = x^2 + c$ has a large enough value for c so that its graph never crosses the diagonal at all. In this case there are no fixed points.

4.13 Composition Machine

Specific Directions. To use the composition machine effectively, care must be taken in plotting the points and drawing the horizontal and vertical segments in the chart. The practice sheet for this activity can be used to establish this geometric process. In going from grid R to grid S, remember that the horizontal segment is extended only to the point where the vertical line from grid P meets it. Your intuition will be hard pressed to accept some of the composition results.

Implicit Discoveries. In moving from grid P to grid R, the output from the first function becomes the input for the second by being reflected off $y = x$ in grid Q. The new output is then plotted in grid S directly over the initial input value used in grid P. Note in question 5 how, for positive values of x, the function $g(x) = x^2$ simply undoes what $f(x) = \sqrt{x}$ does. The result is the diagonal $y = x$.

The composition machine gives a vital piece of information that, at first glance, might be missed. When we see $f(f(x))$, the composition of the function $f(x)$ with itself, look for where it crosses the diagonal $y = x$. Two points will locate the fixed points for $f(x)$. If two other intersection points exist, they will locate exactly the period-two points in $f(x)$. Thus, finding $f(f(x))$ on the composition machine tells us exactly where to draw the box for $f(x)$ about which the spiraling of period-two iteration occurs.

4.1 STAIRCASE AND SPIRAL   **4.1A**

Imagine the path of a point that starts on the x-axis
and repeatedly moves back and forth between a curve
and the diagonal by following these steps:

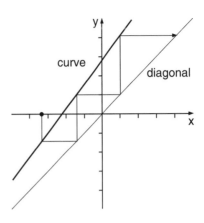

1. Move up or down to the curve.
2. Move right or left to the diagonal.

To create such a path, first draw a vertical segment to
the curve and a horizontal segment from there to the
diagonal. The central geometric process in this activity
is the repeated sequencing of these two steps over
and over again, each time using the last end point as
the next starting point.

This process is called *graphical iteration*. In this activity, we consider only curves that
are straight lines. They are graphs of affine linear functions $y = mx + b$. In the
following activities, graphical iteration will become the central tool used to discover
and discuss the properties of *chaos*.

The intersection of the curve with the diagonal is a special point. Consider starting
the graphical iteration process right to such a point. The process cannot proceed
because moving right or left to the diagonal would not move the point at all. It is
already on the diagonal. Thus, such intersection points are called *fixed points*.

Many different paths can be generated by graphical iteration. Some paths look like
staircases. Those that *staircase in* lead toward a fixed point that behaves as an
attractor. Those that *staircase out* lead away from a fixed point that behaves as a
repeller.

a. Staircase in

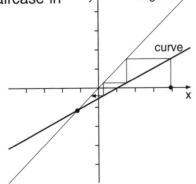

b. Staircase out

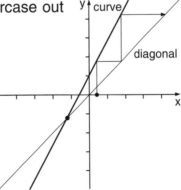

1. Trace each iteration path shown above from the starting point on the x-axis. Does
    it step in toward a fixed point or away from one?

2. Give the approximate coordinates for the intersection point of the curve and the diagonal for each graphical iteration shown on the previous page. Circle the identity of the fixed point as an attractor or a repeller.

    a. $x =$ _____ , $y =$ _____   attractor or repeller
    b. $x =$ _____ , $y =$ _____   attractor or repeller

Use graphical iteration to draw a path from each initial value marked. Indicate if the intersection point of the curve and the diagonal is an attractor or a repeller.

3.
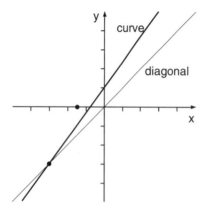

4.
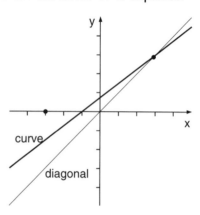

Some graphical iteration paths look like spirals. Those that *spiral in* lead toward fixed points that behave as *attractors*. Those that *spiral out* lead away from fixed points that behave as *repellers*.

a.
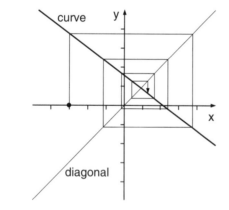

b.
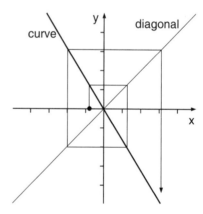

5. Trace each iteration path shown above from the starting point. Does it spiral in toward a fixed point or out and away from one?

6. Give the approximate coordinates for the intersection point of the curve and the diagonal for each figure shown above. Circle the identity of that point as an attractor or a repeller.

    a. $x =$ _____ , $y =$ _____   attractor or repeller
    b. $x =$ _____ , $y =$ _____   attractor or repeller

4.1C

Use graphical iteration to draw a path from each initial value marked. Indicate if the intersection point of the curve and the diagonal is an attractor or a repeller.

7.

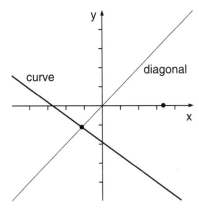

8.

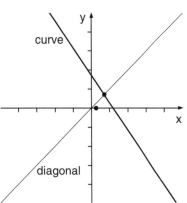

Study each curve in relation to the diagonal $y = x$. Without drawing the path, determine if graphical iteration from the initial value shown will produce a staircase or a spiral. Will the critical intersection point serve as an attractor or a repeller?

9.

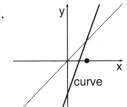

10.

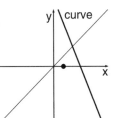

11.

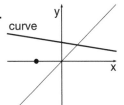

12.

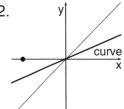

Graphical iteration is always performed using the diagonal $y = x$. In this way, the y-coordinate of every turning point on the curve becomes the x-coordinate of the next turning point on the curve. The x-coordinates of these successive points of iteration on the curve are called the *iterates*.

Complete the list of iterates for the first four turning points on each curve.

13. Staircase in

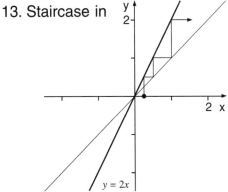

$1/4 \rightarrow 1/2 \rightarrow$ ___ $\rightarrow$ ___

14. Staircase out

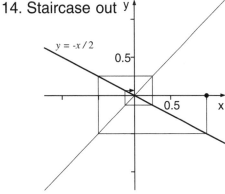

$1 \rightarrow -1/2 \rightarrow$ ___ $\rightarrow$ ___

15. Spiral out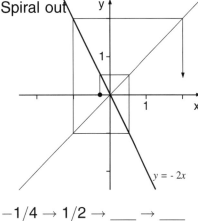

$-1/4 \rightarrow 1/2 \rightarrow \underline{\hspace{1cm}} \rightarrow \underline{\hspace{1cm}}$

16. Staircase in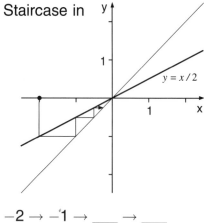

$-2 \rightarrow -1 \rightarrow \underline{\hspace{1cm}} \rightarrow \underline{\hspace{1cm}}$

When a straight line intersects the diagonal $y = x$, the behavior of the graphical iteration around the point of intersection is immediately established.

17. How does the location of the starting point on the axis affect the behavior of the iteration?

18. How does the slope of the straight line affect the behavior of the iteration?

The slope of the straight line used to iterate against the diagonal determines the characteristics of the resulting path.

Slope $m < -1$	Slope $-1 < m < 0$	Slope $0 < m < 1$	Slope $1 < m$
repel & spiral out	attract & spiral in	attract & staircase in	repel & staircase out

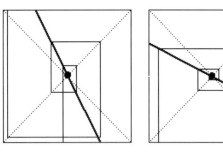 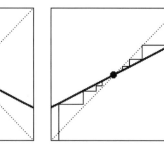 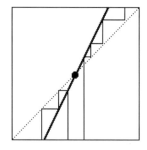

Describe how the graphical iteration behavior changes as the slope of the line

19. increases from just below to just above -1.

20. increases from just below to just above 0.

21. increases from just below to just above 1.

Describe the special behavior of the graphical iteration when the slope of the line

22. is exactly 1.   23. is exactly -1.

4.2 COMPOSITION OF FUNCTIONS   4.2A

The process known as the *composition of functions* is one of the most basic algebraic concepts upon which the study of iteration and chaos is built.

> *Composition of Functions* The formation of a composite function $f(g(x))$ is one function g followed by another f.

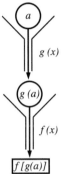

To execute $f(g(x))$, start with the argument a and obtain the functional value $g(a)$. Then use $g(a)$ as the next argument and obtain the functional value $f(g(a))$.

The composition of functions can be represented visually by a diagram such as the one at the right. It can be thought of as a mechanical process in which the $g(x)$ machine processes raw material, a, and produces a product, $g(a)$. Then, the $f(x)$ machine processes that product, $g(a)$, and produces another product, $f(g(a))$.

Use the input given to determine the output for each of the compositions shown. Use the output for $g(x)$ as the input for $f(x)$.

1. 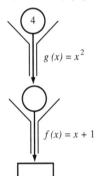   2. 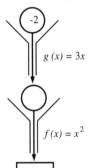   3. 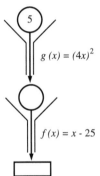   4.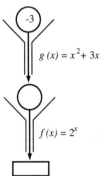

Set up your own diagram and determine the value of $f(g(x))$ for each pair of functions. Use $x = 1$ as the first input value.

5. $g(x) = x + 3$ 6. $g(x) = x^2$ 7. $g(x) = 2x + 5$
    $f(x) = x^2$ $f(x) = x + 3$ $f(x) = 0.5(x - 5)$

Use the functions in questions 5–7. In each case, start with the initial value $x = -1$ and determine the value of the composite function given.

8. $g(f(x))$ 9. $g(g(x))$ 10. $f(f(x))$ 11. $f(g(x))$

12. Given the functions $f(x) = x + 1$ and $g(x) = x^2$, express each of the following compositions in terms of an initial value a.
    a. $f(g(a))$ b. $g(f(a))$ c. $g(g(a))$ d. $f(f(a))$

4.2B

The process of *iteration* is the result of doing something over and over again, repeatedly.

Iteration as a Composition of Functions

Iteration is the process of repeatedly forming the composition of a function with itself. This is illustrated in the diagram at the right in which an argument is repeatedly passed through the same function $f(x)$ in a cyclic fashion resulting in a sequence of functional values as shown below.

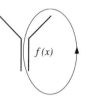

$$x_0 \to f(x_0) \to f(f(x_0)) \to f(f(f(x_0))) \to \cdots$$

Another way of thinking about iteration is to consider some initial argument x_0 and use it to obtain the functional value $f(x_0) = x_1$. Then use x_1 as the next argument and obtain the next functional value $f(x_1) = x_2$, and so on. This iteration of some initial value x_0 for the function $f(x)$ is illustrated in the margin at the right. The resulting *sequence of iterates* or arguments are shown below.

$$x_0 \to x_1 \to x_2 \to x_3 \to x_4 \to x_5 \to \cdots$$

Iterate each x_0 in $f(x) = \sqrt{x}$. Use your calculator and record the results in the table provided, rounded to 3 digits of accuracy.

13. $x_0 = 0.2$ 14. $x_0 = 0.85$ 15. $x_0 = 64$ 16. $x_0 = 1850$

13.	14.	15.	16.
$x_1 = $ _____	$x_1 = $ _____	$x_1 = $ _____	$x_1 = $ _____
$x_2 = $ _____	$x_2 = $ _____	$x_2 = $ _____	$x_2 = $ _____
$x_3 = $ _____	$x_3 = $ _____	$x_3 = $ _____	$x_3 = $ _____
$x_4 = $ _____	$x_4 = $ _____	$x_4 = $ _____	$x_4 = $ _____
$x_5 = $ _____	$x_5 = $ _____	$x_5 = $ _____	$x_5 = $ _____
$x_6 = $ _____	$x_6 = $ _____	$x_6 = $ _____	$x_6 = $ _____
$x_7 = $ _____	$x_7 = $ _____	$x_7 = $ _____	$x_7 = $ _____
$x_8 = $ _____	$x_8 = $ _____	$x_8 = $ _____	$x_8 = $ _____

17. What values do the sequences of iterates in the questions above appear to be approaching? Continue the iteration process to check your conjectures.

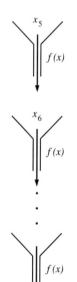

4.2C

18. Use your calculator to iterate the function $f(x) = x^2$. Start with the value x_0 and perform the iteration five times. Write down the resulting sequence of iterates.
    a. $x_0 = 1$ b. $x_0 = 5$ c. $x_0 = 0.2$ d. $x_0 = -3$

Describe the behavior of the sequence of iterates that result from iterating in the function $f(x) = x^2$

19. any initial value $x_0 > 1$.

20. any initial value $0 < x_0 < 1$.

21. any initial value $-1 < x_0 < 0$.

22. any initial value $x_0 < -1$.

Find the function that, under iteration, produces the sequence of iterates listed.

23. $4 \to 5 \to 6 \to 7 \to 8 \to \cdots$ $f(x) = $ _____

24. $1 \to 2 \to 4 \to 8 \to 16 \to \cdots$ $f(x) = $ _____

25. $1 \to -3 \to 9 \to -27 \to 81 \to \cdots$ $f(x) = $ _____

26. $-6 \to -1 \to 4 \to 9 \to 14 \to \cdots$ $f(x) = $ _____

27. $8 \to 4 \to 2 \to 1 \to 1/2 \to \cdots$ $f(x) = $ _____

Iterating a function often produces a sequence of iterates that form an arithmetic or geometric sequence.

Arithmetic sequences   are of the form $a, a + d, a + 2d, a + 3d, ...$, where a is the first term and d is the common difference between successive terms.

Geometric sequences   are of the form $a, ar, ar^2, ar^3, ...$, where a is the first term and r is the common ratio between successive terms.

28. Identify each of the sequences in questions 23–27 as either an arithmetic sequence or a geometric sequence.

Repeated application of f, as in

$$x_0 \to f(x_0) \to f(f(x_0)) \to f(f(f(x_0))) \to \cdots ,$$

soon leads to a notational problem with too many parentheses. We therefore introduce a bit of new notation for the composition of a function f with itself.

$$f^2(x_0) = f(f(x_0)), \quad f^3(x_0) = f(f(f(x_0))), \quad f^4(x_0) = f(f(f(f(x_0)))), \quad ...$$

4.3 GRAPHICAL COMPOSITION AND ITERATION   4.3A

The composition of functions is basic to the study of iteration and chaos. We now explore this process graphically using graphs of the functions involved. The evaluation of a function has been shown by a funnel, capturing the idea of transferring some input x to a transformed output $f(x)$. Here is a graphical representation of such an input-output relationship.

Evaluate First, draw a vertical line from an input value a on the x-axis up to the graph of the function. The intersection point, marked A, has a y-coordinate $f(a)$.

Transfer From A, draw a horizontal line to the intersection point B on the diagonal.

Reflect From B, draw a vertical line up to the top. The x-coordinate at this point is the output value $f(a)$.

Use each initial value given on the x-axis to carry out this process on the function drawn.

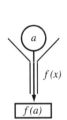
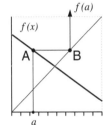
1.
2.
3.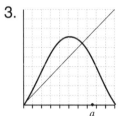

The composition of two functions has been shown by two funnels, the output of the upper one becoming the input of the lower one. Using the graphical representation just given, the composition of two functions can be found in a similar way by stacking one function graph on top of another. Note that the process of evaluating the composition now works in the upward direction. For a given input value a, first determine $g(a)$ with the lower function graph. Then feed the result, as a new input value, into the upper function graph to determine $f(g(a))$.

Carry out the composition of these pairs of functions by this process.

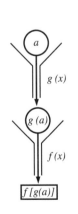
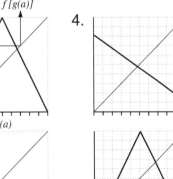
4.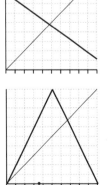
5.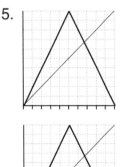
6.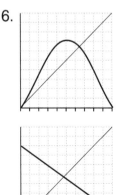
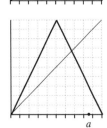

The composition of more than two functions is found in essentially the same way. Stack the functions to compose and then evaluate from bottom to top. Carry out this composition process for each stack of these functions shown.

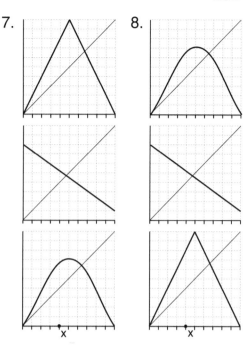

Iteration

The repeated composition of the same functions is called iteration. For two iteration steps, a stack of two identical functions is needed. For three steps, a stack of three is needed. In general, for n steps of iteration a stack of n identical functions is needed.

$$x_0 \rightarrow x_1 = f(x_0)$$
$$x_1 \rightarrow x_2 = f(x_1) = f^2(x_0)$$
$$x_2 \rightarrow x_3 = f(x_2) = f^3(x_0)$$
$$\vdots$$
$$x_{n-1} \rightarrow x_n = f(x_{n-1}) = f^n(x_0)$$

There is a convenient way to perform iteration that avoids stacking a large number of identical function graphs on top of each other. Compress all graphs into a single one and perform all steps of the iteration on this same graph. Graphical iteration of a given initial value x_0 now becomes the repetition of these two steps:

Evaluate   At value x, draw a vertical line to the graph of the function. The y-coordinate of the intersection is $f(x)$.

Feedback   From the intersection point, draw a horizontal line to the diagonal $y = x$. This locates the new x-value.

9.   Trace the initial point x_0 through each stack shown below. Name the iterate located at the top of each stack.

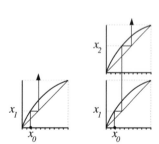

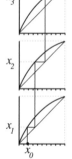

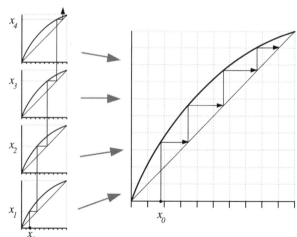

10.   Trace the same initial point x_0 under the single graph shown at the right. Mark the locations of $x_1, x_2, x_3,$ and x_4 on both axes.

4.4 FIXED POINTS AS ATTRACTORS AND REPELLERS   4.4A

Graphical iteration will be used to study the nature of fixed points for linear and nonlinear functions. A *fixed point* is a point where the graph of the function and the diagonal intersect. Therefore, if the x-coordinate of this intersection point is x^*, then the y-coordinate of the intersection point is x^* as well.

Fixed points can have one of several different kinds of behavior with respect to iteration, when we consider initial values close to the fixed point. They can be *attractive* or *repelling*, where accordingly the fixed points are called *attractors* or *repellers*. The characterization of a fixed point can be linked to the *slope* of the graph of the function at that fixed point. If the slope at the fixed point is less than 1 in magnitude, the fixed point is an attractor.

Attractive fixed points come in two kinds. For the first kind (left), we observe a *staircase in* behavior towards the fixed point for all initial values close to the fixed point. The slope of the curve at the fixed point is positive. For the second kind (right), we observe a *spiral in* behavior towards the fixed point for all initial values close to the fixed point. The slope of the curve is negative.

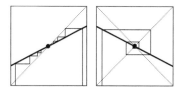

If the slope at the fixed point is greater than 1 in magnitude, the fixed point is a repeller. Repelling fixed points also come in two kinds. For the first kind (left), we observe a *staircase out* behavior away from the fixed point for all initial values close to the fixed point. The slope at the fixed point is positive. For the second kind (right), we observe a *spiral out* behavior away from the fixed point for all initial values close to the fixed point. The slope is negative.

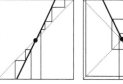

1. Carry out several steps of graphical iteration for the initial values shown. Describe the resulting paths as staircase in or staircase out.

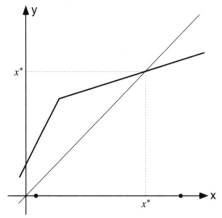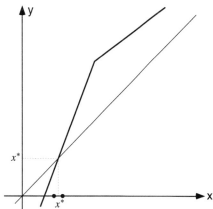

4.4B

2. Carry out several steps of graphical iteration for the initial values shown. Describe the resulting paths as staircase in or staircase out.

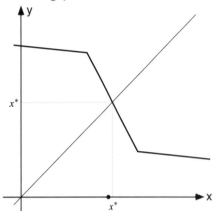   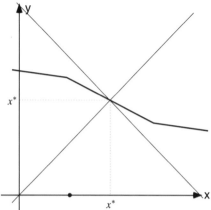

Changing the curve can dramatically alter the path of graphical iteration as it moves back and forth between the curve and the diagonal $y = x$. For example, the iteration path may be confined to an interval, or it may grow beyond all bounds escaping to positive or negative infinity. Two piecewise linear curves are shown below. Draw the corresponding graphical iteration patterns from the starting points shown. In each case, describe the behavior of the iteration path.

3. $y = -2|x| + 2$ 4. $y = -3|x| + 4$

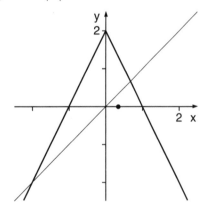   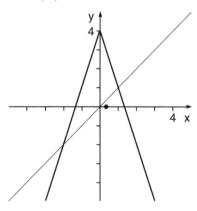

Complete the list of iterates, computing them from the formulas.

5. $x \to -2|x| + 2$: $1/3 \to$ ___ $\to$ ___ $\to$ ___ $\to$ ___ $\to$ ___ $\to$ ___
    What is the long term behavior of the iterates?

6. $x \to -3|x| + 4$: $1/3 \to$ ___ $\to$ ___ $\to$ ___ $\to$ ___ $\to$ ___ $\to$ ___
    What are all the following iterates?

7. Do the graphical iterations from questions 3 and 4 agree in behavior with the computed results in questions 5 and 6?

4.5 SOLVING EQUATIONS AND FINDING PREIMAGES   4.5A

The algebraic composition of functions offers some interesting problems. Given a function f and a range value y, find the value or values of the initial input x such that $y = f(x)$. There may be no, one, two, or more solutions to this problem. Each solution is called a *preimage* of y.

1. When $f(x) = x^2 + 3$ and a value y is given, then $y = x^2 + 3$ is a quadratic equation for the unknown variable x and can be solved. For what values of y does the equation $y = x^2 + 3$ have

    a. two solutions?   b. one solution?   c. no solution?

2. Given the functions $f(x) = x^2 + 3$ and $g(x) = x^2 - 1$, find the value of the initial input x for each of the following compositions. Be careful. Solving for the initial value requires that you reverse the iteration process. For example, to find x in $g(f(x)) = z$ first solve $z = g(y)$ for y, and then $y = f(x)$ for x. Do not consider any intermediate or end results that are negative.

    a. $f(g(x)) = 67$ b. $g(f(x)) = 143$ c. $g(g(x)) = 575$ d. $f(f(x)) = 12$

3. Consider the functions $f(x) = x^2 + 3$, $g(x) = x^2 - 1$, and $h(x) = x^3$. Find the two possible initial values of x if $f(g(h(x))) = 3$.

Graphical composition of preimages

Finding preimages graphically when the graph of the function is given is presented in these examples:

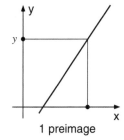
1 preimage

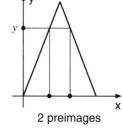
2 preimages

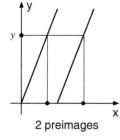
2 preimages

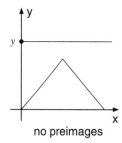
no preimages

4. Find the preimages for the y-values indicated.

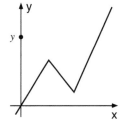

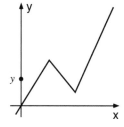

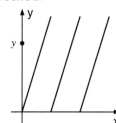

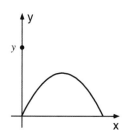

4.5B

The technique introduced on the previous page is the graphical interpretation of solving the equation $y = f(x)$ for x. Now we want to solve equations like $z = g(f(x))$ for x, when z is given. Algebraically, we have learned to do this in two steps. First, solve $z = g(y)$ for y, and then $y = f(x)$ for x.

Graphically, mark z on the upper box, draw a segment down to the diagonal, then a horizontal segment to the graph of g. There may be no, one, two, or more intersection points. From each of these intersection points, draw segments down to the diagonal in the box for f, and then horizontal segments to the graph of f, followed by vertical segments from these intersection points to the lower box.

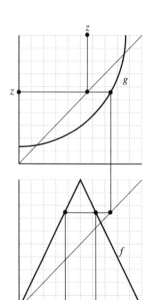
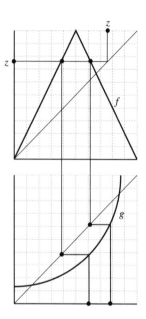

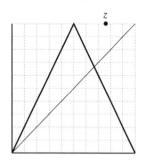
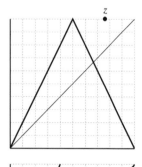

Find the composed preimages for the following functions of f and g.

5.

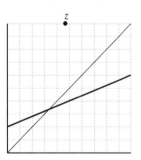

6.

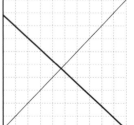

7.

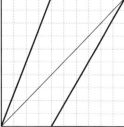

4.6 INTERVALS AND ERRORS   4.6A

Graphical iteration is a mechanical process. When performed with pencil and paper, small errors can be introduced in locating the initial value x_0. This activity investigates the effect graphical iteration has on a small interval around x_0 and the nature of the resulting errors that might be generated when the iterated function is linear.

Consider two new points, b and c, a small but equal distance to the left and to the right of the initial value x_0. Rather than observing the behavior of the point x_0 through successive stages of iteration, we now focus on the corresponding initial interval from b to c for which x_0 is the midpoint. What happens to this interval upon iteration?

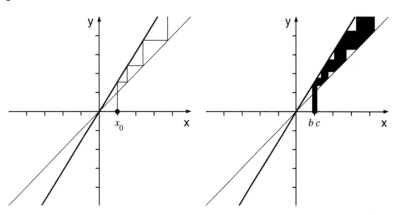

1. Start with the function $f(x) = 2x$. Iteration of x_0 shows a staircase, stepping away from a repelling point, the intersection of the function and the diagonal $y = x$. Describe the corresponding behavior of the interval (b, c) upon graphical iteration.

2. Is the interval expanding, compressing, or remaining constant through iteration?

Suppose $x_0 = 0.20$ with a maximum initial absolute error of $e_0 = 0.01$ either way. This means the actual location of the initial value could be as low as $b = x_0 - e_0 = 0.19$ or as high as $c = x_0 + e_0 = 0.21$. These values locate the boundary points of the error interval, (b, c).

3. Do numerical iteration with the function $f(x) = 2x$ to complete the first five stages of iteration for the interval boundary points. List the corresponding iterates found.

	Initial	Stage 1	Stage 2	Stage 3	Stage 4	Stage 5
Point b	0.19	0.38	____	____	____	____
Point x	0.20	0.40	0.80	1.60	3.20	6.40
Point c	0.21	0.42	____	____	____	____

4. The width of the initial error interval is $0.21 - 0.19 = 0.02$ and the initial, absolute error is half that or $0.02/2 = 0.01$. Complete the list of interval widths and absolute errors through the first five iterations.

	Initial	Stage 1	Stage 2	Stage 3	Stage 4	Stage 5
Interval width	0.02	0.04	____	____	____	____
Absolute error	0.01	0.02	____	____	____	____

149,240

4.6B

5. Describe the behavior of the absolute error as the number of iterations increases.

Relative error is the ratio of the absolute error e_i to the corresponding expected value x_i at a given stage i of iteration. At stage 2, the relative error is 5%.

$$\frac{e_2}{x_2} = \frac{0.04}{0.80} = 0.05 = 5\%$$

6. Compute the relative errors at stages 3, 4, and 5. What appears to be the behavior of the relative errors as the number of iterations becomes large?

For this type of linear function, the interval width is determined and predictable.

7. What is the interval width at stage n? the absolute error? the relative error?

Not all linear functions have intervals and errors that behave the same way. For the function $f(x) = -(1/2)x$, the iterates spiral in toward an attractor. Since the x_i values of the iteration pattern alternate between positive and negative, we will use their absolute values when computing the relative error.

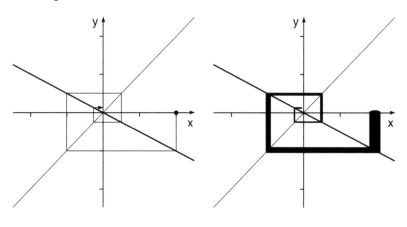

8. Describe the behavior of the error interval upon graphical iteration. Is the interval expanding, compressing, or remaining constant?

9. Suppose the initial value is $x_0 = 1$ with an initial absolute error of $e_0 = 0.02$. Find the interval width, the absolute error, and the relative error for the fifth iterate x_5.

10. Describe the behavior of the absolute error as the number of graphical iterations increases. What happens to the relative error?

Examine these functions. Without performing graphical iteration on the interval shown, determine if it expands, compresses, or remains constant. Then describe how the absolute and the relative errors behave as the number of iterations increases.

11. 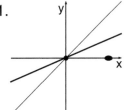   12. 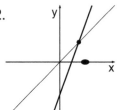   13. 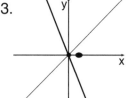   14.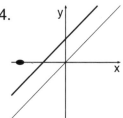

4.7 ITERATING $f(x) = ax(1-x)$: ATTRACTORS   4.7A

Certain classes of functions are of special interest in studying graphical iteration. One such set comes from those functions having the form $f(x) = ax(1-x)$. The coefficient a is a parameter that determines the specific function from within that class.

Consider first the function $f(x) = x(1-x)$ where the parameter takes on the value $a = 1$. Here the diagonal $y = x$ is tangent to the parabola, touching at the single point (0,0). The graph at the right shows that the graphical iteration of the initial value $x_0 = 0.3$ moves in staircase fashion, stepping in toward the origin that serves as an attractor.

1. Trace the graphical iteration path when $f(x) = x(1-x)$ and $x_0 = 0.3$. Describe its behavior.

2. Trace the graphical iteration path when $x_0 = 0.7$. Compare it to that for $x_0 = 0.3$?

3. Will every initial value between 0 and 1 staircase in to the attractor (0,0)?

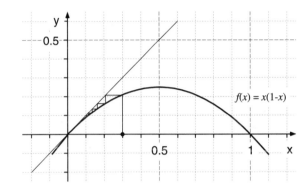

4. Does interval compression or expansion occur under graphical iteration within the interval $[0, 1]$?

When $a = 1.6$, the vertex of the parabola $f(x) = 1.6x(1-x)$ is higher. The diagonal $y = x$ intersects the curve at two different points, (0,0) and (3/8,3/8). The graphical iteration is shown at the right for the initial value $x_0 = 0.2$.

5. Does graphical iteration from $x_0 = 0.2$ show staircase or spiral behavior?

6. The origin is a repelling fixed point. Give the coordinates of the other fixed point and identify it as an attractor or a repeller.

7. In the neighborhood of the two fixed points, is there interval compression or expansion?

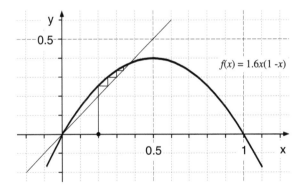

The parabola $f(x) = 1.6x(1-x)$ exhibits an attractive, fixed point under graphical iteration. It is the fixed point (3/8,3/8). Almost all initial values $0 < x_0 < 1$ produce paths that staircase to that point as an attractor. There are two exceptions, namely the fixed point $x_0 = 3/8$ itself, and the initial value $x_0 = 5/8$ which after one iteration terminates in the fixed point.

4.7B

8. Use the graphical method to determine preimages to find the exceptional point x_0 such that the graphical iteration started at x_0 terminates in the fixed point (3/8, 3/8) after one iteration.

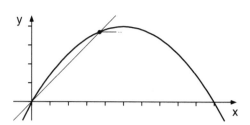

9. Graphical iteration has been performed here on the function $f(x) = 2.8x(1-x)$. The iteration begins with the initial value of $x_0 = 0.10$. Carefully read the first seven iterates from the graph. Record the points, correct to two decimal places. Describe the iteration behavior shown. Is it a staircase or a spiral?

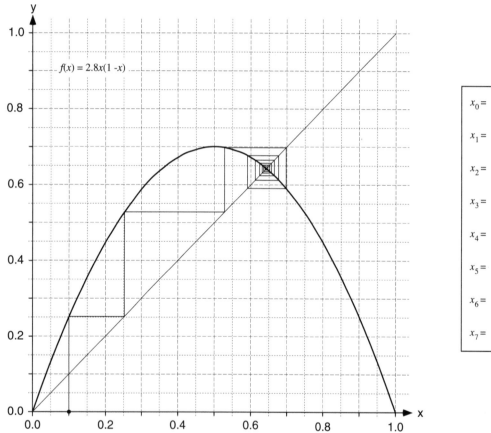

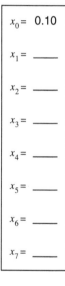

$x_0 = 0.10$

$x_1 = $ _____

$x_2 = $ _____

$x_3 = $ _____

$x_4 = $ _____

$x_5 = $ _____

$x_6 = $ _____

$x_7 = $ _____

10. The curve and the diagonal intersect at the origin. Find the coordinates of the other intersection point. Is this second intersection point an attracting or a repelling point? Is there interval compression or expansion around that point?

11. Perform another graphical iteration on the function $f(x) = 2.8x(1-x)$. Use the same graph above but start with an initial value of your own choice. Describe the iteration behavior and compare it with the description for questions 9 and 10.

4.7C

12. There are iteration paths that do not spiral in to the fixed point! Besides the fixed point itself, there is one initial value that iterates to the fixed point in only one iteration. There are two values that do this in two, and another two values that do it in three iterations. Use the graphical method of finding preimages to determine these initial values.

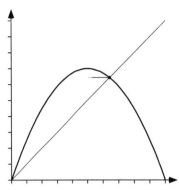

13. Explore the iteration behavior with initial values $x_0 = 0$ and $x_0 = 1$. Speculate on the iteration behavior of values outside the closed interval from 0 through 1.

Surprisingly different results can occur for values of a only slightly greater than 2.8.

14. Graphical iteration has been performed here on the function $f(x) = 3.2x(1 - x)$. Again, the iteration begins with the initial value, $x_0 = 0.10$. Carefully read the first seven successive iterates from this graph. Record the points, correct to two decimal places, in the table to the right of the graph.

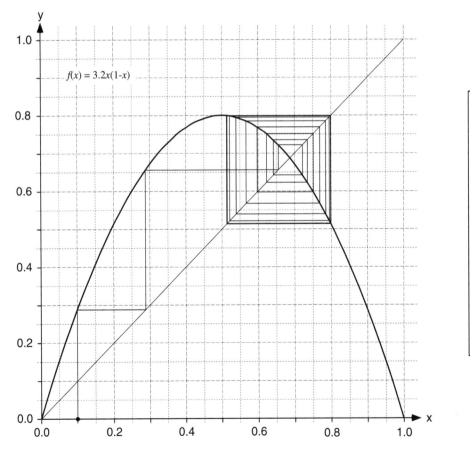

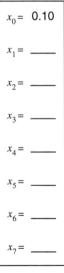

$x_0 = 0.10$

$x_1 = $ _____

$x_2 = $ _____

$x_3 = $ _____

$x_4 = $ _____

$x_5 = $ _____

$x_6 = $ _____

$x_7 = $ _____

4.7D

15. Compare the iterates entered in this table with those found in question 9. Are they essentially alike or are they different?

16. Describe the iteration behavior shown in the graph. Is it a staircase or a spiral? Is there a single attracting or repelling point? Is interval compression or expansion occurring? Or is there a different behavior occurring?

In the graphical iteration shown for $f(x) = 3.2x(1 - x)$, the path appears to approach a square box in a clockwise spiraling and outward motion. Two corners of the box are on the diagonal while the other two are located on the graph of the function. This graphical iteration displays a new periodic or cyclic behavior. The iterates move back and forth, approaching two separate x-values. The approximate values for these two points can be read from the graph. They identify a cycle of period 2, called a *2-cycle*, $x_a \to x_b \to x_a \to x_b \to x_a \to x_b \to \cdots$ Thus, x_a and x_b satisfy the equations $x_b = f(x_a) = 3.2x_a(1 - x_a)$ and $x_a = f(x_b) = 3.2x_b(1 - x_b)$.

Use the same graph of $f(x) = 3.2x(1 - x)$ to complete these questions.

17. Draw on the graph another graphical iteration starting with $x_0 = 0.35$.

18. Is the graphical iteration behavior for $x_0 = 0.35$ essentially the same as that for the initial value of $x_0 = 0.1$? Does the path of the graphical iteration still appear to converge to the same square box as before? Is it spiraling inward or spiraling outward to that box? Is there still interval compression?

19. How do the graphical iterations for initial values $x_0 = 0.65$ and 0.35 compare?

Describe the graphical iteration for each of the following initial values.

20. 0.6875   21. 1   22. below 0   23. above 1

24. How many initial values are there that iterate into the repelling fixed point inside the square box in exactly one, two, or three iterations? Use the graphical method to find preimages to determine these initial values.

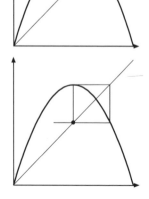

25. In the parabola $f(x) = 2.8x(1 - x)$, use the graphical method to find preimages to determine initial values that iterate exactly into the square box in 1, 2, or 3 iterations.

4.8 ITERATING $f(x) = ax(1-x)$: CHAOS   4.8A

Graphical iteration on the quadratic function $f(x) = ax(1-x)$ can produce various behaviors depending upon the value of the parameter a. This activity focuses on the special case where $a = 4$ and the resulting function is $f(x) = 4x(1-x)$.

1. Perform graphical iteration on the function $f(x) = 4x(1-x)$ shown below. Begin with the initial value $x_0 = 0.10$ and draw the first seven steps in the iteration process. Draw your vertical and horizontal lines as accurately as possible.

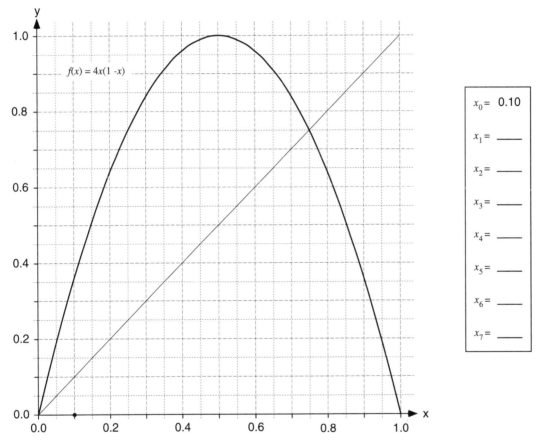

$$x_0 = 0.10$$
$$x_1 = \underline{\quad}$$
$$x_2 = \underline{\quad}$$
$$x_3 = \underline{\quad}$$
$$x_4 = \underline{\quad}$$
$$x_5 = \underline{\quad}$$
$$x_6 = \underline{\quad}$$
$$x_7 = \underline{\quad}$$

2. Read the successive iteration points from the curve, correct to two decimal places. Record them in the table to the right of the curve.

3. How would you describe the behavior of your graphical iteration? Is it a staircase, stepping to an attractor or away from a repeller? Is it spiraling around a fixed point or boxing in a pair of periodic points? Or can you detect some other special type of behavior or pattern in the first seven successive iterates listed?

It is interesting to compare the results with those obtained by other students. For example a table of the various outcomes of the last computed value x_7 will reveal a surprise. The results will *not* be more or less identical with small deviations but rather *distributed throughout the entire unit interval.* This is a consequence of the *sensitivity* to small errors in the iteration of the quadratic function $4x(1-x)$.

Before any final conclusions are drawn from the results above, try graphical iteration again on the same function, $f(x) = 4x(1 - x)$.

4. Choose one initial value x_0 from 0.2, 0.3, and 0.4 and repeat the graphical iteration on $f(x) = 4x(1 - x)$. Record the values in the table.

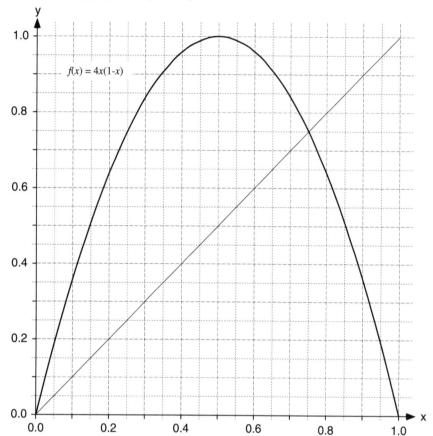

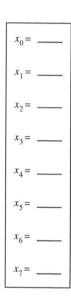

5. Does the behavior of your graphical iteration here appear to be stepping on a staircase, spiraling, or boxing, or is it more chaotic in nature?

Most likely, the graphical iterations completed above illustrate unpredictable, chaotic behavior in which no special pattern appears. However, there are initial values that lead to a simple, non-chaotic behaviors when iterating $f(x) = 4x(1 - x)$.

Describe the sequence of iterates on $f(x) = 4x(1 - x)$ for each of these initial values.

6. 0.50   7. 0.75   8. 1.00   9. 0.25

Many initial values, $0 \le x_0 \le 1$, lead to chaotic behavior when applied, through graphical iteration, to the function $f(x) = 4x(1 - x)$. However, there are initial values in the interval that result in non-chaotic behavior. The point 0.75 immediately iterates to itself as does 0.00. They are the fixed points. Others, such as 0.25, 0.50, and 1.00, reach those same points through iteration.

4.9 GRAPHICAL ITERATION THROUGH TECHNOLOGY   **4.9A**

The graphing calculator provides a quick way to graphically iterate the family of quadratic functions $f(x) = ax(1-x)$. It allows for a convenient study of the short and long term behavior of iterations.

Line	CASIO	Line	TEXAS INSTRUMENTS
1		1	:ClrDraw
2	Fix 3	2	:Fix 3
3	Range 0, 1, 1, 0, 1, 1	3	:0 → Xmin
4		4	:1 → Xmax
5		5	:0 → Ymin
6		6	:1 → Ymax
7	"A="? → A	7	:Disp "A="
8		8	:Input A
9	"I="? → I	9	:Disp "I="
10		10	:Input I
11	Graph Y=AX(1−X)	11	:DrawF AX(1−X)
12	Graph Y=X	12	:DrawF X
13	0 → N	13	:0 → N
14	0 → K	14	:0 → K
15	K=0=>Goto 2	15	:If K = 0
16		16	:Goto 2
17	Lbl 1	17	:Lbl 1
18	AI−AII → I	18	:AI−AII → I
19	N+1 → N	19	:N+1 → N
20	N<K=>Goto 1	20	:If N < K
21	Plot I,I	21	:Goto 1
22	Goto 3	22	:Goto 3
23	Lbl 2	23	:Lbl 2
24	Plot I,0	24	:Line(I, 0, I, I)
25	Lbl 3	25	:Lbl 3
26	AI−AII → J	26	:AI−AII → J
27	Plot I,J	27	:Line(I, I, I, J)
28	Line	28	:Line(I, J, J, J)
29	Plot J,J:Line △	29	:Pause
30	J △	30	:Disp J
31		31	:Pause
32	J → I	32	:J → I
33	N+1 → N	33	:N+1 → N
34	N<K+12=>Goto 3	34	:If N<K+12
35		35	:Goto 3
36		36	:End

Explanation:

Lines:	1–6	Initialize graphics and set the domain and range of the viewing window.
	7–12	Initialize the parameter a, the initial value x_0, and draw the function and diagonal.
	14–22	Carry out a number of K iterations without display.
	25–35	provide 12 iterations with graphics and display of the current x-value.

4.9B

Follow these steps upon execution of either program listed:

- Enter a, the parameter value or coefficient for the function $f(x) = ax(1 - x)$.
- Enter the initial value x_0.
- Press ENTER or EXE.
- After each iteration step, press ENTER or EXE again to continue the process.

Up to this point, we have seen several different graphical iteration behaviors for the parabola $f(x) = ax(1 - x)$, all determined by the initial value and the parameter a. In the following exercises we will reconfirm our observations and extend the range of phenomena found in the iteration of quadratic functions.

1. Enter the program in the calculator and run it with the parameters and initial values explored in Activities 4.7 and 4.8. Compare the results with the graphs of the earlier activities. For $a = 1.6$ start with $x_0 = 0.2$, and for $a = 2.8, 3.2, 4.0$ with $x_0 = 0.1$.

2. Use the program, with initial value $x_0 = 0.1$, to decide what the iteration behavior is for each for the given parameter. Mark the corresponding cell in the table.

$a =$	1.50	2.90	3.24	3.90
Chaos				
Period-2 attractor				
Fixed point, spiral in				
Fixed point, staircase in				

For many parameters a, the first few iterations do not begin to reveal the final behavior of the iteration. In this case, it may help to perform some number of pre-iterations before actually plotting the iteration path. This will dampen the transient effect of a particular choice of initial condition.

3. Change line 13 of the program to allow for K $= 100$ pre-iterations before plotting. Run the program with $a = 3.5$ and initial values $x_0 = 0.2, 0.3, 0.4$. Describe the results and give the x-values of the periodic attractor.

4. Use the program again with K $= 100$ pre-iterations. Test the given values of a and mark the cells corresponding to the long term dynamical behaviors shown.

$a =$	2.95	3.05	3.50	3.68	3.74	3.80	3.84
Chaos							
Period-5 attractor							
Period-4 attractor							
Period-3 attractor							
Period-2 attractor							
Fixed point							

This activity shows that the transition from order to chaos, as a increases to 4, is not a simple one. Initially, there is a period doubling sequence with the period increasing from 1 to 2 to 4 and, in fact, to 8, 16, ... But for larger parameters, attractors of diverse periods seem to alternate with chaos. This topic is continued in Activity 5.13.

4.10 EXPANSION AND COMPRESSION IN $f(x) = 4x(1-x)$ **4.10A**

We have seen many behavioral patterns from graphical iteration applied to functions of the form $f(x) = ax(1-x)$. For some values of a, these parabolas produce staircases while for other values, spirals about a fixed point or spirals approaching a periodic cycle of iteration occur. Yet other parameters such as $a = 4$ lead to chaos and unpredictable behavior. This activity focuses on why these differences occur.

We investigated earlier the effect of interval iteration on linear functions. How does interval iteration behave on functions that are not linear? At first, the behaviors will seem little different from the staircases, spirals, attractors, repellers, compression, and expansion seen earlier. However, with the quadratic $f(x) = 4x(1-x)$, there will be some special surprises.

Perform numerical iteration on each function with the given initial values. Use your calculator to find the first, second, and third iterates in each case.

1. $f(x) = \sqrt{x}$ using $x_0 = 2.5$ and $x_0 = 3.0$.

2. $f(x) = x^2$ using $x_0 = 1.1$ and $x_0 = 1.2$.

3. For each of the two functions above, find the differences between the respective iterates at the computed stages. Then determine if iteration tends to expand or compress the initial difference that exists between the two initial values.

The same two functions are graphed below. On each graph, perform a few steps of graphical iteration using the same two initial values as in exercises 1 and 2. Compare the geometric results found here with the numerical ones found in question 3.

4.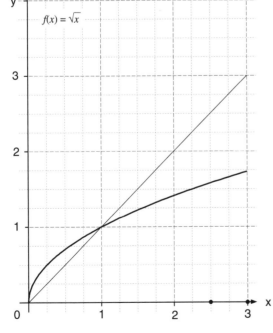

5.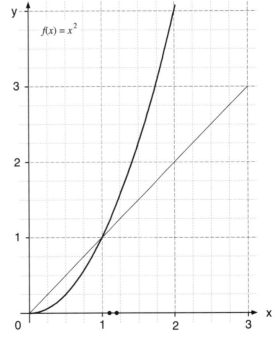

4.10B

Two different types of behavior appear in the last two questions.

- For $f(x) = \sqrt{x}$, the interval compresses as the staircase steps in to the attractor.

- For $f(x) = x^2$, the interval expands as the staircase steps out from the repeller.

In the parabola $f(x) = 4x(1-x)$, these two different iteration patterns of compression and expansion are combined. Of particular interest are the transition points from one behavior to another. Their locations occur when the slope of the curve is 1 and -1.

The slope of a curve at a given point is defined by the slope of the tangent line to the curve at that point. For the parabola $f(x) = 4x(1-x)$, the slope changes incrementally from a large positive value to a large negative value.

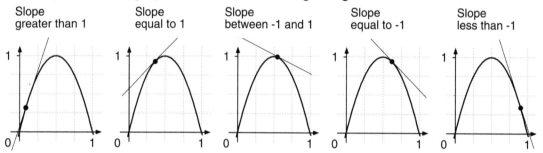

| Slope greater than 1 | Slope equal to 1 | Slope between -1 and 1 | Slope equal to -1 | Slope less than -1 |

The points on the curve where the slopes are 1 and -1 can be found using the square grid on which the parabola is drawn. These tangent lines form 45 degree angles with the horizontal and vertical lines of the grid. They are parallel to the diagonals of the squares in the grid and can easily be found using a ruler.

There are two points on the graph of $f(x) = 4x(1-x)$ with slopes of 1 and -1. They have been marked at $x = 3/8$ and $x = 5/8$.

6. Identify all intervals in the domain of the graph of the function $f(x) = 4x(1-x)$ for which one step of graphical iteration will expand an interval.

7. Identify all intervals for which one step of graphical iteration will compress an interval.

8. Write a rule relating compression and expansion to the slope of the function.

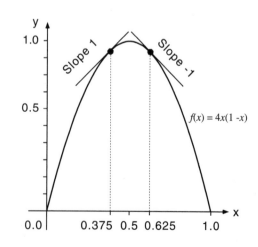

4.10C

The graphs shown below are magnifications of those graphs shown in questions 4 and 5. Continue the iteration patterns initiated on both graphs until you can determine if the pattern forms a staircase in towards an attracting fixed point or out away from a repelling fixed point.

9.

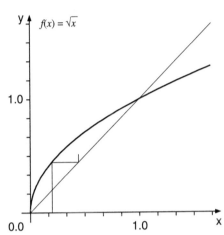

10.

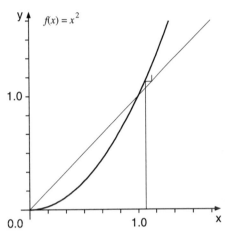

11. Write a rule relating the tendency for graphical iteration to follow a staircase in or out to the properties of compression and expansion.

For each function, identify the intervals in the domain $[0, 1]$ for which graphical iteration compresses intervals. Points where the slopes are 1 or -1 are marked.

12.

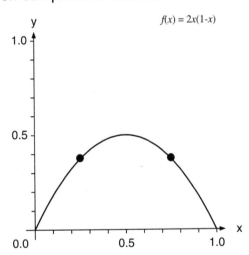

13.

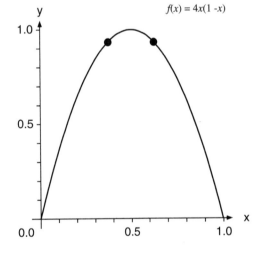

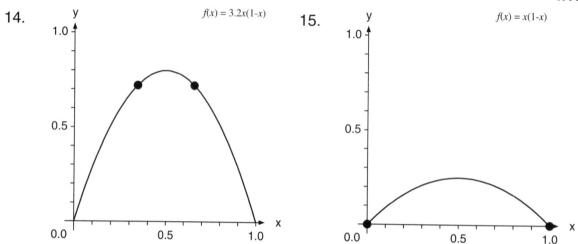

14. $f(x) = 3.2x(1-x)$
15. $f(x) = x(1-x)$

16. Based upon the relative size of the intervals found in questions 12–15, for which function will inaccuracy during iteration lead most frequently to error expansion rather than error reduction?

Interval expansion and compression are determined within the neighborhood of a point by the slope of the curve at that point.

• Expansion occurs when the slope is less than -1 or greater than 1.

• Compression occurs when the slope is between -1 and 1.

• The rate of expansion and compression is determined by the extent to which the slope differs from these values of -1 and 1.

Curves are most steep where their slopes are most extreme, well below -1 or well above 1. In these cases, small drawing errors most quickly expand into large ones.

Graphical iteration is a repeated drawing process where successive iterates are read from points on the graph. Many functions produce iteration behaviors that are stable and predictable. However, as the slope becomes more steep, error expansion around those points becomes more dramatic. In the case of $f(x) = 4x(1 - x)$, only a very small portion of the interval from 0 to 1 has a slope between -1 and 1 where error compression occurs. Once iteration has arrived in that portion, already in the next step it will be in the steep, expanding portions again creating major error expansions. The result is an intermittent variation between error expansion and compression, leading to wildly divergent behavior over the entire interval.

Thus, the small drawing errors in the function $f(x) = ax(1 - x)$ will be most severe when the parameter a is the greatest. In order for the sequence of iterates to stay within the range, $0 \le f(x) \le 1$, the maximum value possible for a is 4. As a approaches 4, this surprising, chaotic phenomenon occurs. We did not see it for smaller values of a such as 1.0, 1.5, 2.0, 2.8, and 3.2.

4.11 SENSITIVITY   **4.11A**

The programmable calculator provides a quick way to numerically iterate the function $f(x) = ax(1 - x)$. In this activity, we use a program to study the issues of predictable and unpredictable behavior in the iteration. We will see how small errors in the graphical or numerical evaluation of the function may or may not lead to chaos. This is a consequence of the presence or absence of *sensitivity*.

The following program performs a total of 7 iterations with initial condition and parameter a provided by the operator. In addition, it allows the user to specify the precision of the calculation. After each iteration, the result is truncated to a specified number F of digits. This provides for the opportunity to simulate and study those errors that typically arise when graphically iterating with paper and pencil.

Line	CASIO	Line	TEXAS INSTRUMENTS
1		1	:ClrHome
2		2	:Disp "ENTER A"
3	"A="? → A	3	:Input A
4		4	:Disp "ENTER X"
5	"X="? → X	5	:Input X
6		6	:Disp "DCML PLCS"
7	"DCML PLCS"? → F	7	:Input F
8	0 → N	8	:0 → N
9	Lbl 1	9	:Lbl 1
10	AX−AX^2 → X	10	:AX−AX^2 → X
11	Int((10 x^y F) X)÷(10 x^y F) → X	11	:IPart ((10∧F)X)/(10∧F) → X
12	N+1 → N	12	:N+1 → N
13	N△	13	:Disp N
14	X△	14	:Disp X
15	" "	15	:Disp " "
16		16	:Pause
17	N<7 => Goto 1	17	:If N<7
18		18	:Goto 1
19		19	:End

Lines: 1–5 Set up display, enter the parameter a and initial value x_0.

6–7 Determine the precision by setting the number of decimal places to be used in the computation.

10–16 Contains the body of the iteration loop that evaluates the quadratic function, truncates the result to the specified number F of digits, and displays the iteration count and the iterate.

Follow these steps upon execution of either program listed:

- Enter the number of decimal places desired.
- Enter a, the parameter value or coefficient for the function $f(x) = ax(1 - x)$.
- Enter the initial value x_0.
- Press ENTER or EXE.
- After each iteration step, press ENTER or EXE again to continue the process.

4.11B

In questions 1–4, set the number of decimal places used in the computation at 5 but round the results to be entered in the tables to 3 decimals.

1. Perform numerical iteration for $f(x) = 2.8x(1 - x)$ using the initial values given.

x_0	x_1	x_2	x_3	x_4	x_5	x_6	x_7
0.100							
0.450							
0.700							

2. Perform numerical iteration for $f(x) = 3.2x(1 - x)$ using the initial values given.

x_0	x_1	x_2	x_3	x_4	x_5	x_6	x_7
0.100							
0.450							
0.700							

3. In what way do the results in the two tables support the behaviors already associated with the graphical iteration of these two functions performed with pencil and paper earlier?

4. Perform numerical iteration for $f(x) = 4x(1 - x)$ using the initial values given.

x_0	x_1	x_2	x_3	x_4	x_5	x_6	x_7
0.100							
0.450							
0.700							

5. Study the results in the table for question 4. In what way do they support the unusual behavior associated with $f(x) = 4x(1 - x)$?

The above calculations have been performed with 5 digits of accuracy. After each iteration the result was truncated to 5 decimal places at line 11 of the program.

4.11C

Now explore iterations that allow for some larger tolerances for error by using only 3 decimal places of accuracy.

6. Set the number of decimal places to 3 and the initial value to 0.1. For each of the parameters $a = 2.8$, 3.2, and 4.0, carry out 7 iterations and record the results.

a	x_0	x_1	x_2	x_3	x_4	x_5	x_6	x_7
2.8	0.100							
3.2	0.100							
4.0	0.100							

7. Compare the results in the table with those obtained in exercises 1–4. For x_7 express the differences bewtween 3-digit and 5-digit computation in terms of %.

While small errors in the computation for parameters $a = 2.8$ and $a = 3.2$ do not affect the final iteration behavior, there are apparent perturbations for $a = 4.0$. The chaotic behavior of graphical iteration on the function $f(x) = 4x(1 - x)$ arises from the magnification of small errors that are created and propagated in each step of the pencil and paper process of drawing. The next exercises are set up to further explore these deviations.

8. Change line 17 of the program to allow for twelve successive iterations.

9. This time, set the number of decimal places at 10. Enter $a = 4$ for the parameter and $x_0 = 0.1$ for the initial value I. Carry out the numerical iteration process 12 times. Then repeat the process using only 3 decimal places of accuracy. Record the results in the table leaving the last column empty.

Iterations	10 decimal places	3 decimal places	difference
0	0.1000000000	0.100	0.000
6			
7			
8			
9			
10			
11			
12			

10. Were the parameter chosen to be $a = 2.8$, the results of the two runs would not differ. However, at $a = 4$ small rounding errors introduced at every iteration do make a lot of difference. By how much do the results differ after 6–12 iterations? Fill out the corresponding entries in the right hand column of the table using three places of accuracy. Describe your findings in words.

4.11D

11. Change line 17 of the program to allow for 30 successive iterations. Repeat the entire process using 10 decimal places of accuracy for both runs. This time, however, use the initial value 0.1000000001 for the second run. Compare the results as before and express the findings in words.

Iterations	10 decimal places	10 decimal places	difference
0	0.1000000000	0.1000000001	0.000
10			
20			
25			
30			

From the computations just completed, it becomes apparent that, when small rounding errors are introduced, they can quickly grow into large errors, even through numerical iteration. This lack of predictability in certain systems was observed in 1961 by Edward Lorenz, an MIT meteorologist, and later became known as *The Butterfly Effect*. The term came from a paper Lorenz presented in 1979 titled "Predictability: Does the Flap of a Butterfly's Wings in Brazil Set Off a Tornado in Texas?"

When a mathematical system is unstable and chaotic, the iteration reveals a behavior described as *sensitive dependence on initial conditions*. In systems that exhibit this behavior, any slight change or perturbation of the initial or intermediate points typically results in dramatically different outcomes.

The function $f(x) = 4x(1 - x)$ exhibits this sensitivity under iteration and that is why its behavior is chaotic when we allow small rounding errors, as was done in question 9. But will that behavior always show up, even when we don't intentionally round back? Does computation on the calculator always produce such results, always generating that slight change or error needed to seed the process that leads to the chaotic behavior? The unfortunate answer to the last question is *yes*.

Errors will eventually be introduced in the iteration process on every calculator and computer, when the iteration behavior is in the expansion mode. And these errors can propagate themselves rapidly. This dramatic effect is the unavoidable consequence of *finite digit arithmetic*. It is inherent in all digital computation from cheap pocket calculators through the multi-million dollar supercomputers. Under the right conditions, it can create unexpected and completely meaningless results!

4.12 ITERATING QUADRATICS   4.12A

Quadratic functions can produce a wide variety of graphical iteration behaviors. Specific characteristics are exhibited around fixed points that serve as attractors and repellers. This activity views iteration behaviors for quadratic functions in two different forms, $f(x) = ax(1 - x)$ and $f(x) = x^2 + c$.

In geometric form, the quadratic function has the shape of a parabola. In algebraic form, the general quadratic function is written in terms of three parameters p, q, and r.

$$f(x) = px^2 + qx + r$$

When $p = -a$, $q = a$, and $r = 0$, the quadratic function can be expressed as $f(x) = ax(1 - x)$, a form explored in detail earlier in this unit. All parabolas in this form intersect the x-axis at (0,0) and (1,0). The parabolas rise higher and higher and the sides become steeper and steeper as the parameter a increases. In each case, the pattern of graphical iteration depends on the value of the parameter a.

When $1 \leq a \leq 4$, all points in $0 \leq x_0 \leq 1$ forever stay within those limits under iteration. The most interesting dynamics of graphical iteration occur in this interval. All points outside the interval escape to negative infinity. Other results occur when $a < 1$ or $a > 4$.

1. Describe the iteration behavior shown for the function $f(x) = 0.5x(1 - x)$ with the initial value $x_0 = 1.5$. Where is the attractor?

2. In the long run, is the iteration behavior for all the points in the interval $1 < x_0 < 2$ essentially the same as that shown? What about the points in the two intervals, $-1 < x_0 < 0$ and $0 < x_0 < 1$? What are the iteration behaviors for the individual points -1, 0, and 1?

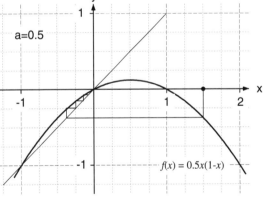

3. Continue the iteration shown for the function $f(x) = 5x(1 - x)$ with the initial value $x_0 = 0.9$. Describe the long term behavior. Is there an attractor or can a point in the interval $0 \leq x_0 \leq 1$ escape to negative infinity?

4. Find a point in $0 \leq x_0 \leq 1$ that does not escape to negative infinity.

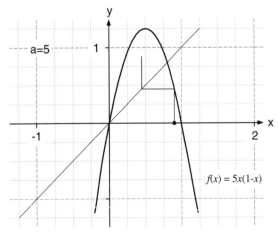

4.12B

Consider again the general quadratic function given in the form $f(x) = px^2 + qx + r$. When $p = 1$, $q = 0$, and $r = c$, the quadratic function is $f(x) = x^2 + c$. All parabolas in this form have vertices on the y-axis at $(0, c)$. As the parameter c increases, the parabola is translated upward. However, the steepness at any given point on the curve does not change as c changes.

5. Let $f(x) = x^2 - 0.65$. Describe the iteration behavior shown for $f(x) = x^2 - 0.65$ when the initial value x_0 is 0.1. Does it appear that the intersection point of the parabola and the diagonal is an attractor for the graphical iteration?

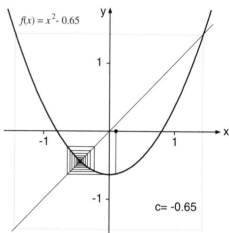

6. Over what interval of initial values x_0 would you expect the iteration behavior to have the same fixed point attractor as that shown? Study the relevance of the square box in dashed lines.

7. Let $f(x) = x^2 - 1$. Describe the iteration behavior shown for $f(x) = x^2 - 1$ when the initial value x_0 is 0.5. What is the attractor and how is it approached?

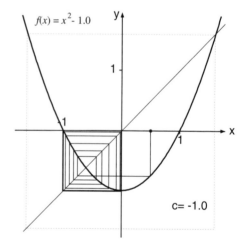

8. Over what interval of initial values x_0 would you expect the iteration behavior to have the same periodic attractor as that shown? Study the relevance of the square box in dashed lines.

9. Let $f(x) = x^2 + 0.35$. Describe the iteration pattern shown for $f(x) = x^2 + 0.35$ when the initial value x_0 is 0.1. Are there any attractors or repellers?

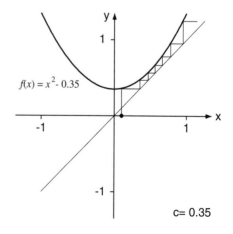

10. In the long run, the iteration pattern is one of a staircase, stepping out to positive infinity. However, the first few steps appear to be moving in. How would you describe the behavior of a small error interval around this iteration path? Would you expect it to be compressing, expanding, both, or neither?

Iteration of $f(x) = x^2 + c$ **4.12C**

Use your graphing calculator and this program to complete the remaining questions. This program is the same as that from Activity 4.10 except that it uses the modified quadratic function $f(x) = x^2 + c$ and it allows the user to enter the range for the graphical display in lines 5 and 6.

Line	CASIO	Line	TEXAS INSTRUMENTS
1		1	:ClrDraw
2	Fix 3	2	:Fix 3
3	"RANGE"? $\rightarrow$ R	3	:Disp "RANGE"
4		4	:Input R
5	Range -R, R, 1, -R, R, 1	5	:-R $\rightarrow$ Xmin
6		6	:R $\rightarrow$ Xmax
7		7	:-R $\rightarrow$ Ymin
8		8	:R $\rightarrow$ Ymax
9	"C="? $\rightarrow$ C	9	:Disp "C="
10		10	:Input C
11	"I="? $\rightarrow$ I	11	:Disp "I="
12		12	:Input I
13	0 $\rightarrow$ N	13	:0 $\rightarrow$ N
14	Graph Y=XX + C	14	:DrawF X$\wedge$2 + C
15	Graph Y=X	15	:DrawF X
16		16	:I$\wedge$2 + C $\rightarrow$ J
17	Plot I,0	17	:Line(I, 0, I, J)
18		18	:Goto 2
19	Lbl 1	19	:Lbl 1
20	II + C $\rightarrow$ J	20	:I$\wedge$2 + C $\rightarrow$ J
21	Plot I,J	21	:Line(I, I, I, J)
22		22	:Lbl 2
23	Line	23	:Line(I, J, J, J)
24	Plot J,J:Line $\triangle$	24	:Pause
25	N+1 $\rightarrow$ N	25	:N+1 $\rightarrow$ N
26	N $\triangle$	26	:Disp N
27	J $\triangle$	27	:Disp J
28		28	:Pause
29	J $\rightarrow$ I	29	:J $\rightarrow$ I
30	N<20=>Goto 1	30	:If N<20
31		31	:Goto 1
32		32	:End

Describe your observations after 20 iterations for each value of c. Watch both the graph and the corresponding numerical iterates. Use 6 decimal places of accuracy and be sure to change the range values in the program as needed.

11. Use $c = 0.40, 0.35, 0.30$, and 0.25, in each case starting with $x_0 = 0.2$.

12. Use $c = -0.6, -0.7, -0.8$, and -0.9, in each case starting with $x_0 = 0.3$.

13. Choose values around $c = -2$ and study the corresponding iteration behavior.

PRACTICE SHEET FOR 4.13

g(f(x))

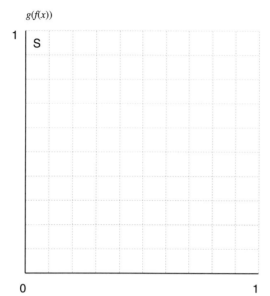

g(x)

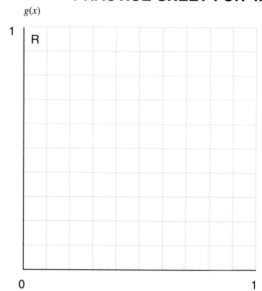

f(x)

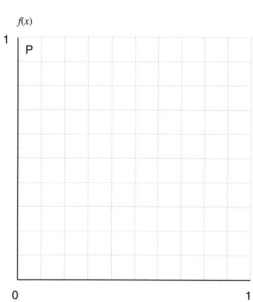

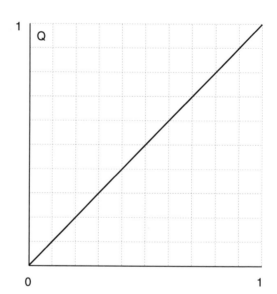

4.13 COMPOSITION MACHINE   4.13A

The composition $g(f(x))$ describes at one time the result of applying the function $g(x)$ on $f(x)$. One approach identifies this composition through algebraic substitution. The composition machine described in this activity is a pencil and paper procedure to construct $g(f(x))$ geometrically. The technology of the graphing calculator offers yet another view of the composition of functions.

The method uses a square array of four grids P, Q, R, and S. The diagonal $y = x$ is always fixed in grid Q. To find $g(f(x))$, draw $f(x)$ in grid P and $g(x)$ in grid R. The composition $g(f(x))$ can then be constructed in grid S in a point-by-point manner.

Step 1: From a point a on the horizontal axis of grid P, draw a vertical line up through grid S.

Step 2: Locate the point $(a, f(a))$ where this line intersects $f(x)$ in grid P. Draw a horizontal line from this point to the diagonal $y = x$ in grid Q.

Step 3: Reflect this line off $y = x$ as a vertical line drawn up to the graph of $g(x)$ in grid R to locate the point $(f(a), g(f(a)))$.

Step 4: From this point, draw a horizontal line to grid S. The point where this line meets the vertical line drawn in Step 1 is the new composition point $(a, g(f(a)))$.

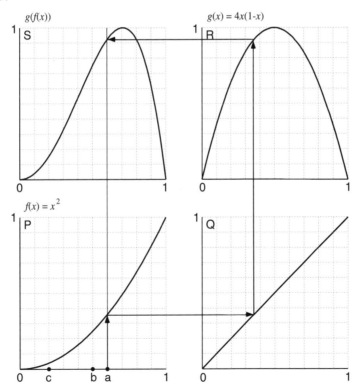

1. On the chart above, follow the path from a through $f(a)$ to $g(f(a))$. Then draw in the paths from b through $f(b)$ to $g(f(b))$ and from c through $f(c)$ to $g(f(c))$.

4.13B

2. In the composition machine chart just shown, we have $f(x) = x^2$ and $g(x) = 4x(1 - x)$. Write the algebraic expression for $g(f(x))$.

We may think of repeatedly performing the steps listed on the previous page as a kind of *composition machine*. As the points in the domain of the function $f(x)$ in grid P pass through the function $g(x)$ in grid R, the composition machine generates $g(f(x))$ in grid S.

In the chart, point a is 0.6. From 0.6, the composition machine generates 0.9216.

$$a \quad \rightarrow \quad f(a) \quad \rightarrow \quad g(f(a))$$
$$0.6 \quad \rightarrow \quad 0.36 \quad \rightarrow \quad 0.9216$$

3. Point b is 0.5. Find the value of $f(0.5)$ and $g(f(0.5))$. Do they agree with the locations of the points drawn in question 1?

4. Point c is 0.2. Find the composition $g(f(0.2))$.

5. In the chart below, we see $f(x) = x^2$ and $g(x) = \sqrt{x}$. Use the composition machine approach with pencil and ruler to sketch the graph of $g(f(x))$. Use the x-values indicated as dots in grid P.

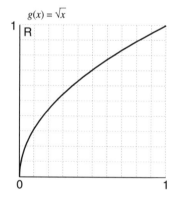

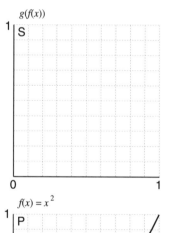

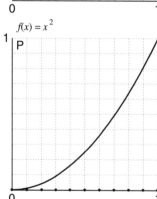

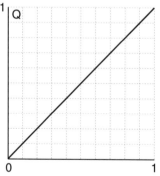

6. Explain why the composition function found in question 5 must be the straight line $g(f(x)) = x$. Then verify it algebraically.

4.13C

The composition machine can be used with two functions $f(x)$ that are the same.
The resulting composition is $f(f(x))$.

7. Use the composition
    machine approach to
    sketch a graph of $f(f(x))$.
    Start with the points 0.00,
    0.25, 0.50, 0.75, and 1.00.
    Remember, the composition
    of two linear functions is a
    linear function.

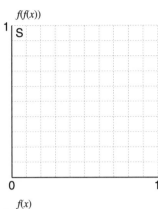

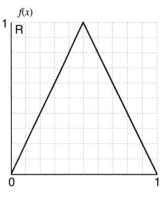

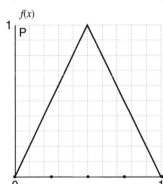

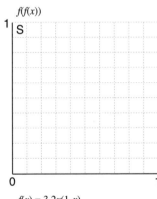

8. The graph of the parabola
    $f(x) = 3.2x(1 - x)$ is
    displayed in the grids P and
    R. Use the composition
    machine approach to
    sketch a graph for $f(f(x))$
    in grid S.

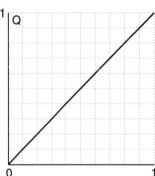

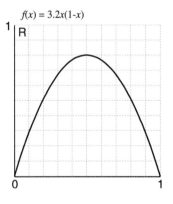

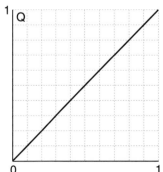

9. Draw the diagonal $y = x$ in the grid S over the composition function $f(f(x))$ completed in question 8. In how many places does the line intersect the curve? Estimate the x-values of the intersection points. Do these points remain fixed when iterating through $f(f(x))$?

10. Use a graphing calculator to graph the composition function drawn in question 8. Compare their shapes. Then draw the diagonal $y = x$ on the function. Use the trace routine to find the x-values of the four intersection points. Compare them to those found in question 9.

Study the graphical iteration of $f(x) = 3.2x(1 - x)$ shown at the left below. There are two obvious fixed points at the intersection points with x-coordinates $x_0 = 0$ and $x_0 = 11/16 = 0.6875$. These turn out to be repellers. There are two other points (0.51304.... and 0.79944...) which form an attractive cycle of period 2 toward which most other points in the interval iterate. It turns out that these four points correspond to the four intersection points on $f(f(x))$, as shown at the right below.

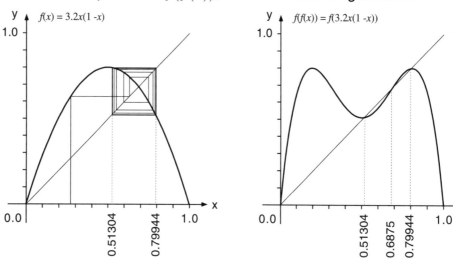

The composition $f(f(x))$ can be used to locate, exactly, the cycle of period two for $f(x)$. It appears as fixed points in the iteration behavior of the composition $f(f(x))$.

$$x_0 = x_2 = x_4 = x_6 = x_8 = x_{10} = x_{12} = \cdots$$

For a range of parameters a in $f(x) = ax(1 - x)$, which includes $a = 3.2$, this cycle of period two is attractive. Each of these points in this attractor appears as a fixed point attractor in the long term iterative behavior of the composition $f(f(x))$.

11. Use your calculator to support algebraically the contention that 0.51304... and 0.79944... are fixed point attractors in $f(f(x))$ when $f(x) = 3.2x(1 - x)$.

Unit 5
Chaos

KEY OBJECTIVES, NOTIONS, and CONNECTIONS

The activities in this unit are designed to develop an understanding of the three main concepts inherent in dynamical systems that exhibit chaos. These concepts are mixing, periodicity, and sensitivity. Systems that are chaotic have all three of these characteristics mingled and merged within them. This is the reason why graphical iteration of different points in a chaotic system produces results that behave in widely different and unpredictable ways.

This unit builds on the intuitive notions of iteration behavior presented earlier. Mixing is first viewed physically through two different processes of kneading dough. Then two corresponding mathematical models are identified, the tent function and the saw-tooth function, and their equivalences to the kneading procedures established. Using binary numbers as a vehicle for investigation, the three important properties of mixing, periodicity, and sensitivity are observed in both functions. The tie is then made to the quadratic function $f(x) = ax(1 - x)$ where similar behaviors exist. Finally, long term behaviors are studied for various values of the parameter a in this quadratic function, first through time series and then through the Feigenbaum plot. This plot, a fractal in its own right, paints a most amazing picture of the journey into chaos.

Connections to the Curriculum

The material covered in these strategic activities forms an integral part of a contemporary mathematics program. The activities may be included in a single unit on this topic or integrated throughout the existing curriculum in those areas to which they are connected.

PRIMARY CONNECTIONS:

Quadratic Functions	Graphing
Evaluating Functions	Piecewise Functions
Numerical Patterns	Visualization
Transformations	Function Composition
Geometric Patterns	Coordinate Geometry
Mappings	

SECONDARY CONNECTIONS:

Sequences and Series	Rational Numbers
Binary Numbers	Irrational Numbers
Linear Functions	Graphing Calculator
Convergence	Limit Concept

Underlying Notions

Chaos

Mathematicians generally agree on three common characteristics of chaos that emerge from the iteration of transformations such as

$ax(1-x)$. These are mixing, periodicity, and sensitive dependence on initial conditions.

Mixing

For any two intervals I and J that are of arbitrarily small but non-zero lengths, there exist points in I that, under iteration, eventually lead to points in J.

Periodicity

Points are periodic if, under iteration, iterates continually recur in a cyclic pattern. Periodic points are abundant in that one can be found near any point in any small subinterval of $[0, 1]$ around that point.

Sensitivity

Small alterations made in the initial position of a point generate dramatically different outcomes in the iterated sequence. This reflects a high degree of sensitive dependence on initial conditions.

The Feigenbaum Plot

A visual way of picturing the nature of the long term behavior of the transformation $ax(1-x)$ under iteration in terms of the different parameter values of a between 1 and 4.

MATHEMATICAL BACKGROUND

The Bigger Picture

Chaos is best understood through the fundamental roles played by its key component parts. These roles are established in the various activities of the unit through illustrations designed to build on the intuitive understanding of the reader and lead into the development of the bigger picture of chaos.

The concept of mixing is described first through the familiar act of kneading dough. The two methods used are stretch-and-fold and stretch-and-cut-and-paste. Initial points are first tracked through these kneading operations and then through graphical iteration under the equivalent piecewise linear tent and saw-tooth functions. In both cases, one can see visually the mixing process in action.

Tent Transformation

Saw-Tooth Transformation

The effects of iteration through both the tent function and the saw-tooth function can be analyzed best using binary numbers. This is because their algebraic definitions involve operations that are particularly easy to execute in base two. For example, doubling a binary number can be done by simply moving the point one place to the right.

Periodicity

One immediate result from doing numerical iteration in binary form for the saw-tooth function is the identification of periodic points. They appear whenever repeating digits occur, and the number of binary digits in the sequence that repeats is the number of iterations needed to complete one cycle in graphical iteration. Since rational numbers occur in repeating digit form, it follows that they exhibit periodic behavior in $[0, 1]$. Some irrational numbers have the property that, under iteration, they visit each and every binary subinterval at every stage in $[0, 1]$. This kind of iterative behavior is called ergodic. Examples can be generated from the labels used in identifying all subintervals at all stages of the binary subdivision.

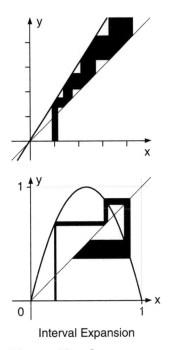

Interval Expansion

Binary Numbers

Mixing

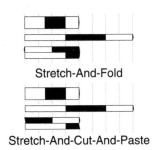

Stretch-And-Fold

Stretch-And-Cut-And-Paste

Sensitivity

The quadratic function in the form of $f(x) = 4x(1-x)$ has behavior virtually identical to the tent and saw-tooth functions since they are equivalent. For example, a particular period-3 cycle under the tent function has a matching one under the parabola. However, the iterate values of the period-3 points under the tent function are transformed under the parabola and occur at other places. In general, periodic points x under T and S correspond to transformed periodic points x' under $f(x)$.

As the parameter a increases toward 4, a larger and larger portion of the entire domain interval $[0,1]$ for $f(x) = ax(1-x)$ produces slopes greater than 1 or less than -1 and these slopes become increasingly steep. This, of course, results in more and more interval expansion. Points become so sensitive to small initial errors that any error whatever explodes into erratic, unpredictable, chaotic behavior as the parameter a approaches 4.

The transition from stable to chaotic behavior as the parameter a moves from 1 to 4 can be seen through separate time series charts for different values of a. To see the entire behavior transition in a single chart, look at the Feigenbaum plot.

Binary numbers help us monitor the iterative behavior of different initial points under the various transformations. Operations with these numbers are introduced because they facilitate the execution of algorithms for the tent and saw-tooth transformations. Binary numbers give a convenient matching between the period of the repeating digits in numerical form and the length of the cycle in the closed path of the iteration of a periodic point in geometric form. Binary numbers are also essential in illustrating the notion of mixing and in developing the arguments for sensitivity since they can be used in the subdivision of the unit interval into subintervals of any size.

An intuitive notion of mixing in a dynamical system can be presented using the analogy of mixing spices in dough through the kneading process. The two different methods presented are both shown to be essentially equivalent to thoroughly mixing the spice. What is so fascinating about this analogy is that it is virtually identical to the iteration process for the tent and saw-tooth functions. The kneading process for stretch-and-fold is equivalent to the tent transformation, while stretch-and-cut-and-paste is equivalent to the saw-tooth transformation. This equivalence is then carried to the quadratic $f(x) = 4x(1-x)$. Mixing occurs through iteration under all of these transformations.

Sensitive dependence on initial conditions is a concept that can be difficult to grasp. It is so counter-intuitive that our sense of how things *should be* blocks our understanding of *what is*. Edward Lorenz, a Massachusetts Institute of Technology meteorologist, had similar difficulty with a weather model he was studying in the early 1960's. It is

LorenzAttractor

The Feigenbaum Map

Feigenbaum Plot

hard for us to understand how, for one function, slight changes in the initial value have no effect on the long term iterative outcomes, while for a very similar function, strikingly disparate results can occur.

Sensitivity is presented in a variety of ways to help the reader gain a sense of this concept. The graphing calculator allows for quickly executed experiments in graphical iteration that can yield both numerical and visual patterns and results. However, the constraints imposed by the finite arithmetic of these machines can be misleading. It is the analysis of sensitivity through the use of binary numbers and binary subdivisions of the unit interval that firmly establish the presence of the property.

The world around us contains many dynamical systems operating together at the same time. The important message to gain from this unit is to be extremely cautious in assuming that all systems have long term behaviors that are predictable.

There are many dynamical systems that have long term behaviors that are chaotic. If we ignore the possibility that such things as global warming, heart beats, fluid flows, and others systems have this potential, we may be in for some catastrophics surprises. This unit gives some mathematical models within our reach that can help us better understand the nature of chaos.

The last activity of the unit explore the long term iterative behavior of the quadratic function $f(x) = ax(1-x)$. The plot used is named after the American physicist, Mitchell J. Feigenbaum, who made this discovery while working at Los Alamos, NM. This Feigenbaum plot gives a phenomenal picture of how the long term iterative behavior patterns of dynamical systems can be dramatically changed by making only slight changes in the value assigned to the parameter a.

The amazing thing about the Feigenbaum plot is its universality. Plots of this type exist for many iterated function models. Through them the understanding of chaos in our world is greatly expanded. So too is our responsibility, both as students and teachers alike, to share this knowledge with others.

Additional Reading

Chapters 10 and 11 of *Fractals for the Classroom, Part Two,* H.-O. Peitgen, H. Jürgens, D. Saupe, Springer-Verlag, New York, 1992.

USING THE ACTIVITY SHEETS

5.1 The Kneading of Dough

Specific Directions. The unit begins with the physical activity of kneading dough, builds on the visualization of the process, and initiates an exploration of the underlying notions of chaos through the phenomena of kneading. Carefully study the stretch-and-fold and the stretch-and-cut-and-paste patterns provided before answering the questions. Pay special attention to the vertical lines drawn in behind each example. These lines are important in helping keep track of where specific sections of dough are transformed through stretch-and-fold and stretch-and-cut-and-paste. Questions 1–15 focus on interval transformations through kneading. The questions from 16 on focus on point transformations through kneading.

Implicit Discoveries. Questions 13–23 lead to the property, established later, that any sequence of n stretch-and-fold iterations are equivalent to $n - 1$ stretch-and-cut-and-paste iterations followed by one stretch-and-fold.

This first activity begins, in an informal way, to focus on the key threads woven throughout this unit on chaos. They are the three properties of chaos: sensitivity, mixing, and periodicity. All three properties operate in the kneading of dough.

5.2 Kneading and Graphical Iteration

Specific Directions. Use questions 1–4 to show how interval iteration on the tent and saw-tooth functions behave exactly as the interval transformations under the kneading operations of stretch-and-fold and stretch-and-cut-and-paste. Use questions 5–10 to show how points are iterated under the two functions graphically and numerically. Be sure to relate how these compare to the mixing of a grain of spice under the two kneading operations. Questions 11–15 show error expansion while the remaining questions tie repeating closed path behavior to function composition.

Implicit Discoveries. Notice how the properties of mixing, sensitivity, and periodicity are woven into the development with the tent and saw-tooth functions. All parts of the tent and saw-tooth functions have slopes that are greater than 1 or less than -1. This implies continuous interval expansion just as repeated kneading operations always stretch the subintervals further and further across the entire interval $[0, 1]$.

5.3 Binary Numbers

Specific Directions. Binary numbers play an important role in the mathematical analysis of the tent and saw-tooth functions and deserve special attention. Those with repeating digits can be expressed as infinite geometric series. Give careful attention to the introduction of binary subdivisions on the unit interval and their number and width at any given stage. This process will play a key role in arguments that follow later.

The algorithm presented on page 5.3C offers a different view through iteration of the conversion process between fractions and decimals, with applications to binary numbers and repeating digits. This is a good place to review the fact that rational numbers can always be written in repeating digit form in any base.

Implicit Discoveries. Every number expressed in repeating digit form, regardless of its base, names a rational number and can be rewritten as an infinite series of fractions. This observation connects the topic directly to that of iteration and geometric series. Rational numbers that have a terminating form in any base also have two distinct repeating-digit forms in that base. Procedures that follow later in the chapter require that all binary numbers be expressed in repeating digit form.

5.4 The Saw-Tooth Function and Binary Numbers

Specific Directions. The implicit discoveries made in Activity 5.3 will help in the computation required for this activity. Stress the two parts of the algorithm for binary evaluation of the saw-tooth function, noting the important role that the first binary digit after the point plays in applying the transformation. Successive iterations are affected by successive binary digits in the expansion.

Use this activity to establish a solid understanding of the notion of labeling binary subdivisions at different stages and how it relates to the initial binary digits that follow the point. This connection will become critical in developing arguments in later sections.

Implicit Discoveries. The saw-tooth transformation involves a shifting of the point in binary numbers between 0 and 1. This moves the location on the unit interval, but that location can always be found at any stage in terms of the leading binary digits. It quickly becomes apparent why analysis through binary numbers is so important.

5.5 The Saw-Tooth Function and Chaos

Specific Directions. Studying the iterative behavior of the saw-tooth function with binary numbers reveals much about mixing, periodicity, and sensitivity. Using binary subdivision labels with the saw-tooth function, we can find, in any subinterval of $[0, 1]$, an initial point that reaches any prescribed target subinterval at any stage. This establishes mixing. Periodic points can be identified in any subinterval at any stage and hence can be as close to a given point as desired. This establishes that periodic points are dense in the unit interval. Building on this idea, points can be generated that begin very close but become a half-unit away through successive iteration under the saw-tooth function. The process includes flipping digits from 0 to 1 or 1 to 0. This duality process will be developed in detail in the next activity. It is used informally here to establish sensitivity, the third property of chaos.

Implicit Discoveries. Repeated binary expansions identify rational numbers and lead to fixed points or periodic cycles. Nonrepeating binary expansions identify irrational numbers. Any listing of binary digits following the point that contains all subdivision labels at all stages generates an irrational number that visits each and every interval at each and every stage.

5.6 The Saw-Tooth and Tent Functions

Specific Directions. This activity begins with a definition of duals with respect to base two and then uses this idea in establishing an algorithm for binary evaluation of the tent function. Stress the two parts of this algorithm, again noting the important role that the first binary digit after the point plays in applying the transformation. Unlike the saw-tooth function, the tent transformation shifts the point and takes duals when the initial binary digit is 1. Successive iterations depend on whether successive binary digits in the expansion are 0 or 1.

Be certain to note how the saw-tooth function comes into play in simplifying repeated tent transformations. The development calls up various techniques established earlier in this unit and the last. The process is explored first with numerical evaluation, by cases in stage two, then by the composition machine, and finally through the stacking of icons. The successive replacement of two T-icons with a T-icon on top of an S-icon illustrates the relationship between these two operators.

Implicit Discoveries. The connection $T^n(x) = T(S^{n-1}(x))$ presented in this activity will be used in the next activity to establish the properties of chaos in the tent function through the knowledge we already have regarding these properties for the saw-tooth function.

5.7 The Tent Function and Chaos

Specific Directions. The properties of chaos, already seen in the kneading of dough, in the iteration of the parabola $f(x) = 4x(1 - x)$, and in the saw-tooth function, now appear in the tent function. The arguments for the existence of mixing, sensitivity, and periodicity come directly from the connection between the saw-tooth and tent functions. The important role of binary numbers in this activity should be emphasized.

Implicit Discoveries. Mixing, sensitivity, and periodicity are all complex notions. The definition of mixing requires only that some point in subinterval I will eventually reach some point in subinterval J through iteration. It does not say every point in I will behave this way nor does it say every point in J can be reached this way. Regarding periodicity, remember that the calculator can only handle rational approximations to irrational numbers. So from a practical point of view, all values entered on the machine will eventually appear as periodic. From a theoretical point of view, should a periodic point be randomly chosen in a subinterval, it will certainly not reach all other subintervals. However, there will always exist a nearby irrational number in the same interval that will. In studying sensitivity, remember that the tent and saw-tooth functions both have slopes that are greater than 1 or less than -1. Hence, interval expansion must always occur upon iteration.

5.8 Population Dynamics

Specific Directions. Use this activity to illustrate the dynamics of iteration in the familiar and disturbing phenomena of population growth and change. It is an excellent example of how mathematicians such as Verhulst search for mathematical models that can be used to describe natural phenomena. The time-series charts illustrate in a new way the notion of long term iteration behavior. The first clearly shows an attractor while the second shows chaotic behavior.

Implicit Discoveries. The Verhulst equation allows for the population fraction P to be greater than 1. This may appear to some to be misleading at first. What the model says is that if the population fraction exceeds 100%, the factor of proportionality r turns from positive to negative and the population stops increasing and begins to decrease.

5.9 The Parabola

Specific Directions. The Verhulst equation for modeling population dynamics is a quadratic function. This activity shows how it is related to the familiar parabola $f(x) = ax(1 - x)$. The population fraction P corresponds to the value x and the proportionality factor r corresponds to the parameter a through the relationship $a = r + 1$. Be careful in making the necessary modifications to the calculator program in Activity 5.8 in order to implement the Verhulst equation.

Implicit Discoveries. The exact mathematical equivalence of x_0 under $f_a(x)$ and $(r+1)/r\, x_0$ under $g_r(x)$ should be supported by the results of questions 9–12. However, we have noted the sensitivity property of this quadratic, and the results of questions 13 and 14 illustrate just how dramatic that effect can be at higher iterates. Small rounding errors introduced by the calculator ultimately destroy completely this mathematical equivalence, rendering both sets of computed iterates useless.

5.10 Signs of Chaos for the Parabola

Specific Directions. The interval values generated in questions 3–7 give the first five stages when iterating the small initial interval $[0.08, 0.10]$. Note where the point 1.00 of stage 4 moves on the fifth iteration. Use the first graph in question 9 to contrast the visual appearances of period-1, period-2, and period-3 points. Use the second graph to illustrate the potential complexity of periodic paths of higher order.

Implicit Discoveries. The function $f(x) = 4x(1-x)$ exhibits chaotic behavior, making predictability an impossibility. It is important to recognize that there are algebraic methods for finding periodic points of any period n. This presence of infinitely many periodic initial points establishes that they are everywhere dense in $[0, 1]$. The difficulty in solving $f^n(x_0) = x_0$ for these periodic points of period n should not cloud this issue.

5.11 Transformation from Tent to Parabola

Specific Directions. Use this activity to establish the equivalence of the tent function and the parabola $f(x) = 4x(1-x)$. Questions 1–4 show graphically how the two functions behave under interval iteration. The tent function expands throughout while the parabola expands around the extremes and compresses around the middle of the unit interval. The trigonometric function given transforms the uniform square grid in such a way as to visually illustrate the equivalence of the tent and parabola, and hence their corresponding iterations as well. Transforming the image of the path of x_0 under $T(x)$ produces the path of the corresponding x'_0 under $f(x)$.

Implicit Discoveries. The equivalence of the tent function to the parabola establishes the behavior for the parabola. The trigonometric proof on 5.11D is complete once mathematical induction is applied. The parabola has chaotic behavior, exhibiting all three properties of sensitivity, mixing, and periodicity.

5.12 Long Term Behavior and Time Series

Specific Directions. This activity emphasizes that the behavior of a very long sequence of iterates obtained from functions of the form $f(x) = ax(1-x)$ depend upon the setting of the parameter a. The essential issue involves a dichotomy between the stable pattern on the one hand and rather chaotic and unpredictable behavior on the other. The difference can best be viewed by performing graphical iteration as shown in question 1 and then plotting time series of the values of the iterates against the number of iterates as in question 2.

Implicit Discoveries. As parameter a increases from 1 to 4, the long term behavior tends to become more erratic.

5.13 The Feigenbaum Plot

Specific Directions. For some settings of parameter a in $f(x) = ax(1-x)$, the sequence of iterates generated by $f(x)$ tends to settle upon a finite set of points that repeatedly occur. Such a finite set of points acts as an attractor for the iterative pattern in the sense that, irrespective of the initial point, the sequence eventually cycles through the values of the attracting set. In this activity, our objective is to plot on a single graph the eventually emerging iteration pattern associated with each setting of the parameter a.

Implicit Discoveries. As parameter a increases from 1 to 3, the long term behavior is stable and highly predictable. As the parameter a increases from 3 to 4, the long term behavior tends to become more erratic

5.1 THE KNEADING OF DOUGH   5.1A

The analysis of chaos requires an understanding of the fundamental properties and concepts of sensitivity, mixing and periodicity. This activity uses the intuitive notion of kneading dough to illustrate them. When we turn to a more mathematical interpretations by geometric and numerical iterations, these terms will become more precise.

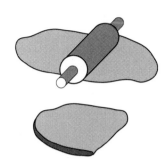

How do we knead dough? We can stretch it using a rolling pin and then fold it. By repeating this over and over again, the spices in the dough can be mixed thoroughly.

In much the same way, points can be mixed within an interval through certain geometric iterations. The more kneading, the greater is the mixing of the spices in the dough. Similarly, the greater the number of geometric iterations, the more thorough the distribution of the points in the interval when mixing occurs.

Consider these two kneading transformations. Watch how the coloring of the interval changes from start to finish in each case. Then compare the two final results.

Stretch-and-fold Stretch to twice the length. Bend at the center.
               Fold the right half up and over onto the left half.

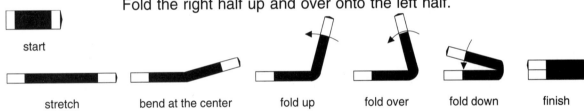

start

stretch   bend at the center   fold up   fold over   fold down   finish

Stretch-and-cut-and-paste Stretch to twice the length. Cut in half. Move the right
               half up, slide it over the left half, and paste it down.

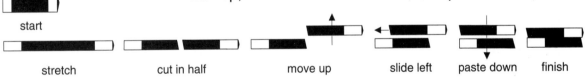

start

stretch   cut in half   move up   slide left   paste down   finish

Both intervals in the illustration above started out colored the same. Yet the finished results from the two transformations are colored differently.

Three-eights of the initial interval width is shaded. Is the shading mixed throughout more of the interval width after the transformation?

1. Stretch-and-fold   2. Stretch-and-cut-and-paste

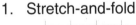

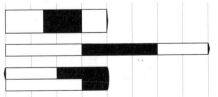   

Shade in the corresponding locations of the marked intervals at the remaining stages to complete these successive stretch-and-fold transformations.

Initial stage

First stretch-and-fold

Second stretch-and-fold

3.

4.

5. Has the color been distributed across the entire interval after each of the two successive stretch-and-fold transformations shown in questions 3 and 4?

Shade in the corresponding locations of the marked intervals at the remaining stages to complete these successive stretch-and-cut-and-paste transformations.

Initial stage

First stretch-and-cut-and-paste

Second stretch-and-cut-and-paste

6.

7.

8. Has the color been distributed across the entire interval after each of the two successive stretch-and-cut-and-paste transformations of question 6 and 7?

Work backwards through the two stretch-and-fold transformations shown, shading in where the corresponding coloring must have been at the first and initial stages.

Initial stage

First stretch-and-fold

Second stretch-and-fold

9.

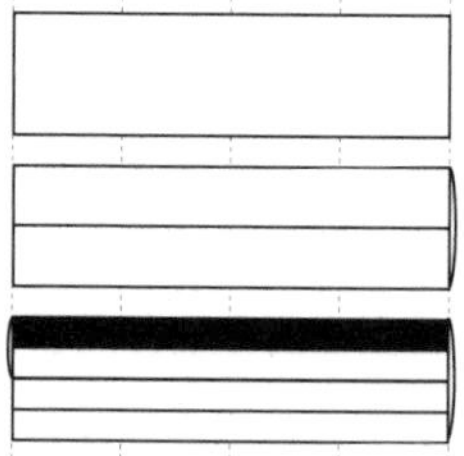

10.

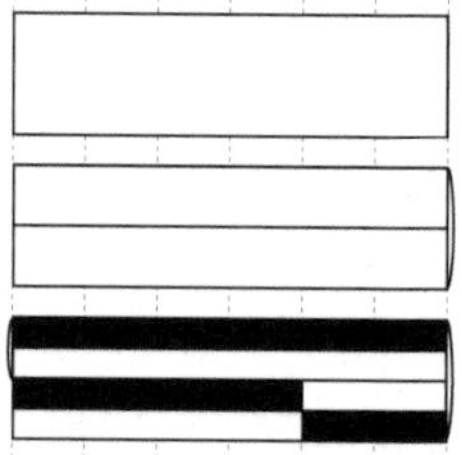

Work backwards through the two stretch-and-cut-and-paste transformations shown, shading in where the corresponding coloring must have been at the earlier stages.

Initial stage

First stretch-and-cut-and-paste

Second stretch-and-cut-and-paste

11.

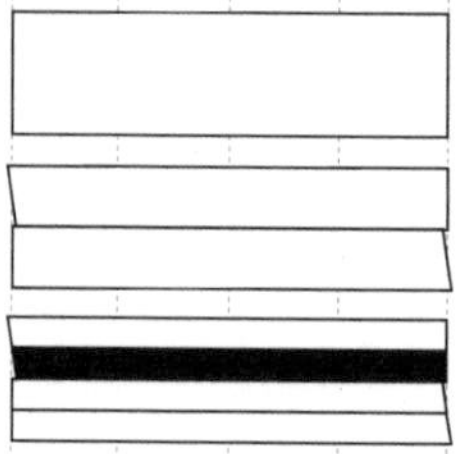

12.

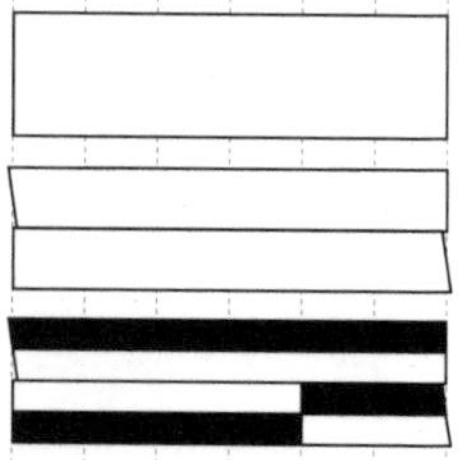

5.1C

Complete these sequences of transformations by entering the letters A through L in the appropriate cells at each successive stage.

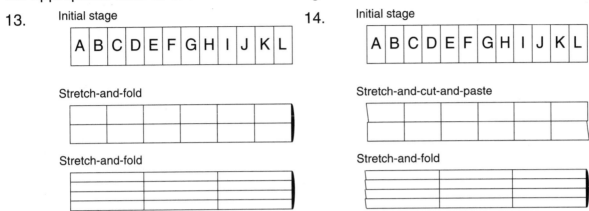

13.   Initial stage   14.   Initial stage

Stretch-and-fold   Stretch-and-cut-and-paste

Stretch-and-fold   Stretch-and-fold

15. Compare the final mixing results in questions 13 and 14. In what ways are they different? In what ways are they the same? Has the same mixing across the interval width occurred in both cases?

The results of questions 13 and 14 illustrate that two stretch-and-fold transformations yield the same result as one stretch-and-cut-and-paste followed by a stretch-and-fold transformation. In both case, the same letters occur in each of the three parts across the interval. An identical mixing of letters has occurred across the two intervals, left to right, even though the letters are stacked differently, top to bottom.

Let us look at this result from a different point of view by tracking the path of a single grain of spice through the kneading process. Ignore the thickness of the dough and think only of the position of the point on the segment.

16. The figures at the left below show the positions of an initial point as it moves through two successive iterations of stretch-and-fold transformations. Mark its location after a third stretch-and-fold transformation. Use the underlying common subdivision to find the position.

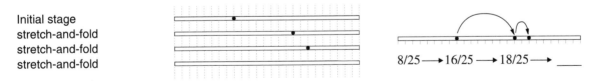

Initial stage
stretch-and-fold
stretch-and-fold
stretch-and-fold

$8/25 \longrightarrow 16/25 \longrightarrow 18/25 \longrightarrow$ ____

17. The figure at the right shows the same positions of the point as an orbit on a single segment. The positions are given as fractions to identify their locations on a unit segment. Extend the orbit to a third stretch-and-fold and give the corresponding numerical value of its final location on the segment.

5.1D

18. The figures at the left below show the positions of an initial point as it moves through two successive iterations of stretch-and-cut-and-paste transformations. Mark its final location if it is followed by a stretch-and-fold transformation.

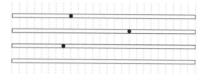

Initial stage
stretch-and-cut-and-paste
stretch-and-cut-and-paste
stretch-and-fold

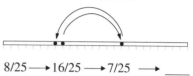

$8/25 \longrightarrow 16/25 \longrightarrow 7/25 \longrightarrow \underline{}$

19. The figure at the right shows the same positions of the point as a path or orbit on a single segment. Extend the orbit to a third transformation that is stretch-and-fold and give the corresponding numerical value of its final location.

20. Are the final locations in questions 16 and 17 the same as in 18 and 19?

As a general rule, we will see that three stretch-and-fold iterations are equivalent to two stretch-and-cut-and-paste transformations followed by one stretch-and-fold.

21. Test this rule with some more examples of your own.

Draw the two orbits for particles at 7/25 and 8/25 and complete the number sequences for five iterations.

22. Stretch-and-fold

$7/25 \rightarrow 14/25 \rightarrow \underline{} \rightarrow \underline{} \rightarrow \underline{} \rightarrow \underline{}$

7/25

$8/25 \rightarrow 16/25 \rightarrow \underline{} \rightarrow \underline{} \rightarrow \underline{} \rightarrow \underline{}$

8/25

23. Stretch-and-cut-and-paste

$7/25 \rightarrow 14/25 \rightarrow \underline{} \rightarrow \underline{} \rightarrow \underline{} \rightarrow \underline{}$

7/25

$8/25 \rightarrow 16/25 \rightarrow \underline{} \rightarrow \underline{} \rightarrow \underline{} \rightarrow \underline{}$

8/25

Observe how the repeated stretching pulls nearby particles more and more apart. This is the property of sensitivity. On the other hand, the folding or pasting can bring these same points back close together, sometimes producing repetitions.

24. Draw the orbit for repeated stretch-and-fold iterations that start at 1/5. Describe the behavior.

$5/25 \rightarrow 10/25 \rightarrow \underline{} \rightarrow \underline{} \rightarrow \underline{} \rightarrow \underline{}$

25. Draw the orbit for repeated stretch-and-cut-and-paste iterations that start at 1/5. Describe the behavior.

$5/25 \rightarrow 10/25 \rightarrow \underline{} \rightarrow \underline{} \rightarrow \underline{} \rightarrow \underline{}$

A particle at 1/5 returns, under stretch-and-cut-and-paste, to its original position after four iterations. After four more iterations, it is back again, and so on. Under stretch-and-fold, a particle at 1/5 moves to 2/5 and then returns after two iterations. These two examples illustrate periodic movement.

5.2 KNEADING AND GRAPHICAL ITERATION   5.2A

Much of mathematics consists of constructing abstract models of natural phenomena. The physical activity of kneading dough can be modeled through mathematical transformations with the same kinds of properties. These transformations are described algebraically and geometrically over the interval $[0, 1]$.

The mathematical model for the stretch-and-fold kneading operation is called a tent transformation because of its shape.

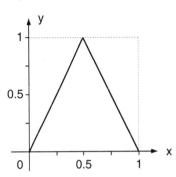

$$T(x) = \begin{cases} 2x & \text{when } 0.0 \leq x < 0.5 \\ 2 - 2x & \text{when } 0.5 \leq x \leq 1.0 \end{cases}$$

We can show the equivalence between the stretch-and-fold of kneading and the tent transformation through these activities.

1. Intervals A, B, and C on the x-axis have been matched with their transformed locations on the y-axis. Complete the tent transformations for intervals D, E, and F. Observe that all intervals are stretched. What is the new size of the interval relative to the original size?

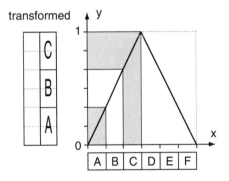

2. Intervals A, B, and C have been transformed through the kneading operation of stretch-and-fold. Complete the entries for intervals D, E, and F and compare them with the results of question 1.

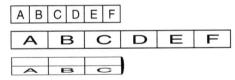

The tent function stretches all intervals of the left and right half of the unit interval to twice their original size and folds back all intervals of the right half, just the way kneading by stretch-and-fold does.

The mathematical model for the stretch-and-cut-and-paste kneading operation is called a saw-tooth transformation because of its shape.

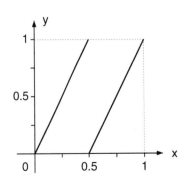

$$S(x) = \begin{cases} 2x & \text{when } 0.0 \leq x \leq 0.5 \\ 2x - 1 & \text{when } 0.5 < x \leq 1.0 \end{cases}$$

We can show the equivalence between the stretch-and-cut-and-paste of kneading and the saw-tooth transformation.

5.2B

3. Intervals G, H, and I on the x-axis have been matched with their transformed locations on the y-axis. Show the saw-tooth transformations for intervals J, K, and L. Observe that all intervals are stretched. What is the new size of the interval relative to the original size?

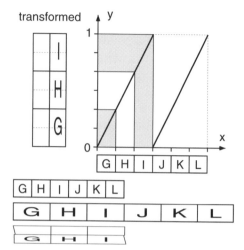

4. Intervals G, H, and I have been transformed through the kneading operation of stretch-and-cut-and-paste. Complete the entries for J, K, and L and compare them with question 3.

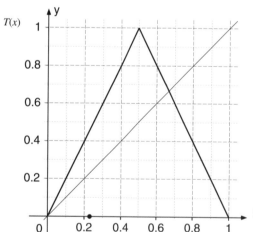

The saw-tooth function stretches all intervals of the left and right half of the unit interval to twice their size. Each half of the unit interval is transformed to the whole interval, and, thus, the stretch-and-cut-and-paste kneading process is represented.

Successive steps of graphical iterations on the saw-tooth or tent function are equivalent to successive steps of kneading by stretch-and-cut-and-paste or stretch-and-fold. Complete the graphical iteration through six stages to show how the marked point x_0 is transformed to x_1, x_2, x_3, x_4, x_5 and x_6 by these functions.

5.

6.

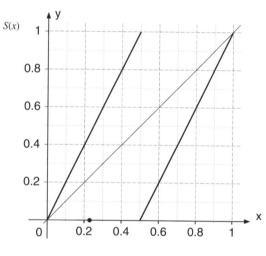

Assume the initial values of exercises 5 and 6 are at 0.23. Compute the first six iterations using the equations of the tent and the saw-tooth functions.

7. Tent function   0.23 → _____ → _____ → _____ → _____ → _____ → _____

8. Saw-tooth function   0.23 → _____ → _____ → _____ → _____ → _____ → _____

Compare the result of the graphical and numerical iterations. They probably do not agree. Perhaps the initial points were not exactly at 0.23! Try the two different initial values 0.22 and 0.24 that are very close.

5.2C

9. Tent function   0.22 → ____ → ____ → ____ → ____ → ____ → ____ → ____

                            0.24 → ____ → ____ → ____ → ____ → ____ → ____ → ____

10. Saw-tooth function   0.22 → ____ → ____ → ____ → ____ → ____ → ____ → ____

                            0.24 → ____ → ____ → ____ → ____ → ____ → ____ → ____

Observe how the initially small differences in question 9 and 10 grow rapidly. This is the property of sensitivity. For graphical iteration, this means that even the smallest drawing errors will be amplified rapidly in the course of the iteration.

Let us now consider a small interval such as $[0.22, 0.24]$ that contains 0.23. Continue the process for two more iterations.

11.
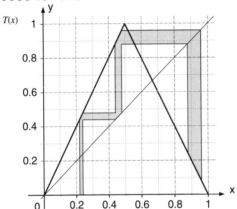

12.
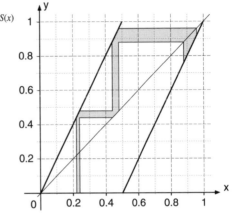

13. After how many iterations will the small interval become mixed throughout the whole unit interval, covering the complete range from 0 to 1?

This same mixing behavior can be found for any interval. The iterates will finally cover the entire range from 0 to 1. This does not mean that when iterated, all points of the interval visit all possible positions of the unit interval. In fact, it is possible that the iterates of a single point behave very differently and visit only a very limited number of points in the unit interval.

14. Describe the behavior of graphical iteration at $x = 2/3$, the intersection point of the diagonal and the graph of the tent function.
    Describe the behavior of a grain of spice under kneading by stretch-and-fold when it starts at the position corresponding to this same intersection point.

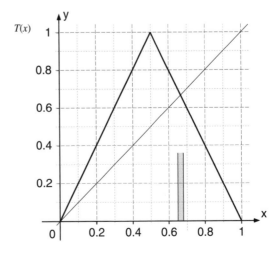

15. What happens to a small interval such as $[0.65, 0.68]$ that contains the fixed point at 2/3? What is its size after three iterates?

5.2D

Compute the first three iterates and draw the corresponding path of graphical iteration.

16. Tent function

2/5 → _____ → _____ → _____

17. Saw-tooth function

1/3 → _____ → _____ → _____

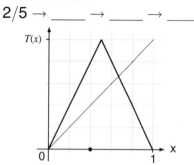
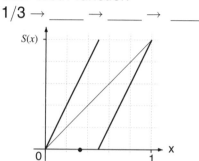

In exercise 16, the iteration oscillates back and forth between 2/5 and 4/5, forming a cycle of period 2. We have $T(T(2/5)) = 2/5$ and $T(T(4/5)) = 4/5$. In other words, both are fixed points of the composition function $T(T(x))$. This function can be determined using the compostition machine of Unit 4. In question 17, we have a period-2 cycle of the saw-tooth function formed by 1/3 and 2/3.

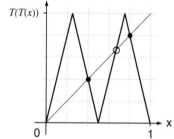

The intersection points of the diagonal and the composite function graphs represent the fixed points of the iteration. The open circle in the left graph marks the original intersection point for T. The remaining marked points correspond exactly to the period-2 cycles for T. A similar relationship exist in the right graph for S.

18. The left figure shows the graph of $S(S(S(x)))$. The marked intersection points correspond to a period-3 cycle of S, such that $S(S(S(x))) = x$.

$$3/7 \rightarrow 6/7 \rightarrow 5/7 \rightarrow 3/7 \rightarrow 6/7 \rightarrow \dots$$

The right figure shows the corresponding path of graphical iteration. Find the second period-3 cycle and draw the second close path.

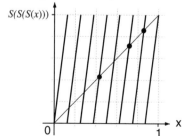
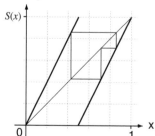

5.3 BINARY NUMBERS   5.3A

What determines the behavior of specific points in $[0, 1]$ that are iterated through the tent and saw-tooth functions? The explanation involves the use of binary notation. This activity focuses on those special properties of binary computation needed.

In order to express values in the decimal system, the digits 0, 1, 2, 3, 4, 5, 6, 7, 8, and 9 are used with place values that are successive powers of ten.

place values	10^2	10^1	10^0	10^{-1}
decimal numeral	7	3	9	5

$739.5_{ten} = 7(100) + 3(10) + 9(1) + 5(1/10)$

The same number 739.5 can be expressed using place values that are powers of some other positive base. In a base-two binary system, only the two digits 0 and 1 are used with place values that are powers of two.

place values	2^9	2^8	2^7	2^6	2^5	2^4	2^3	2^2	2^1	2^0	2^{-1}
binary numeral	1	0	1	1	1	0	0	0	1	1	1

$1011100011.1_{two} = 1(512) + 0(256) + 1(128) + 1(64) + 1(32) + 0(16) + 0(8) + 0(4) + 1(2) + 1(1) + 1(1/2)$

1. Complete the first five powers of three expressed in base two.

    11   1001   11011   _____   _____

Fractions with powers of two as denominators are easy to express in binary form.

$$3/4 = 1/2 + 1/4 = 0.1_{two} + 0.01_{two} = 0.11_{two}$$

2. Complete the first five powers of three-fourths expressed in base two.

    0.11   0.1001   0.011011   _____   _____

Multiplying a decimal number by its base value ten is equivalent to moving the decimal point one place to the right. Similarly, multiplying a binary number by its base value two can be accomplished by moving the point one place to the right.

Compare the product of this binary multiplication by two with doubling by addition in base two.

$10 \times 10111.001 = 101110.01$ $\begin{array}{r} 10111.001 \\ +10111.001 \\ \hline 101110.010 \end{array}$

Multiplying a number by two is equivalent to adding the number to itself. Verify this relationship by first multiplying the binary number by two and second, by adding the binary number to itself. In each case, the two results should be the same.

3. 11.101   4. 101.1   5. 0.01101

Digits can repeat in binary form just as they can in decimal form.

$$4/7 = 0.5\overline{714285714285}_{ten} = 0.1\overline{00100}_{two}$$

Every number expressed in repeating or eventually repeating digit form in any base can be written as an infinite series of fractions. These are infinite geometric series with the constant multiplier from one term to the next given as the ratio r.

$$0.666\overline{6}_{ten} = 6/10 + 6/100 + 6/1000 + 6/10,000 + \cdots \quad r = 1/10$$
$$0.10\overline{1010}_{two} = 1/2 + 1/8 + 1/32 + 1/128 + \cdots \qquad\qquad r = 1/4$$

If $-1 < r < 1$, then the sum S of the infinite geometric series can be found by

$$S = a/(1 - r),$$

where a is the first term and r is the ratio of successive terms.

In the above examples we have $a = 0.6_{ten} = 6/10$ and $a = 0.10_{two} = 1/2$. Write each of the following numbers as an infinite geometric series. Find the value for r and use the formula for the sum to find the value of the series.

6. $0.25\overline{25}_{ten}$ 7. $0.00\overline{1}_{two}$ 8. $0.\overline{1010}_{two}$

As a consequence, we note the following interesting relationship:

$$\overset{n \text{ zeros}}{0.0...00\overline{1}_{two}} = (1/2)^n = \overset{n - 1 \text{ zeros}}{0.0...01_{two}}$$

If we divide the unit interval into two halves, each binary number of the left half starts with the digit 0 after the point, while each binary number of the right half starts with the digit 1.

9.   Mark the starting binary digits that belong to each of the remaining subinterval as the process of creating a binary subdivision by halving the intervals of the previous stage is repeated.

First stage

| | 0.0... | | 0.1... | |
|0| | 1/2 | | 1|

Second stage

| | 0.01... | | | |
|0| | 1/2 | | 1|

Third stage

|0| | | 1/2 | | 1|

10. The first two digits after the point of the binary numbers that belong to an interval of the stage-2 subdivision are the same. How many are the same for stage 3?

11. How many subintervals are obtained if we continue the process to stage 4, stage 5, and stage 6? What is the length of a single interval of the subdivision of stage 4, stage 5, and stage 6?

At stage n of the binary subdivision of the unit interval, we obtain 2^n subintervals of length $(1/2)^n$. Each of the subintervals collect those binary numbers that have the same n starting digits after the point. In other words, binary numbers that have the first n digits in common have at most a distance of $(1/2)^n$ between them.

5.3C

Converting a fraction into its repeating digit form requires a small algorithm that reveals the digits step by step. For example, the repeating decimal form of a number can be obtained by repeatedly

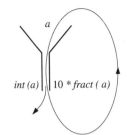

- subtracting the integer part, int(a)
- and then multiplying by 10 the remaining fractional part, fract(a).

When recording the integer parts provided by this algorithm, we obtain all decimals of a given number. Let us find the decimal representation of 4/7.

12. First, we obtain the fractional part 4/7 and the integer part 0. Multiplying the fractional part by 10 provides 40/7. Now the fractional part becomes 5/7 and the integer part is 5, and so on. Complete the table and verify that

$$4/7 = 0.\overline{571428}.$$

	input a	fract (a)	int (a)
	4 / 7	4 / 7	0
x 10	40 / 7	5 / 7	5
x 10	50 / 7	_____	_____
x 10	10 / 7	_____	_____
x 10	_____	_____	_____
x 10	_____	_____	_____
x 10	_____	4 / 7	8

Note that the integer parts repeat as soon as the fractional parts repeat. Use the algorithm to express each fraction in its repeating decimal form.

13. 6/7

14. 3/11

15. 5/13

Most pocket calculators use more digits for the computations than for the display. How can we reveal the hidden digits for the number π, displayed as 3.141592654. Here is the trick: first, we subtract the integer part 3 and then multiply the result by 10. The result, 1.415926536, uncovers one additional digit. The repetition of this simple procedure is nothing else but our algorithm to convert fractions into the decimal form.

16. If you apply the algorithm to the repeating-decimal representation of 4/7, you can see why the algorithm works. Watch how the decimals are shifted to the left until they become visible as integer parts.
    Complete the table and compare with exercise 12.

	input a	fract (a)	int (a)
	0.$\overline{571428}$	0.$\overline{571428}$	0
x 10	5.$\overline{714285}$	0.$\overline{714285}$	5
x 10	7.$\overline{142857}$	_____	_____
x 10	1.$\overline{428571}$	_____	_____
x 10	4.$\overline{285714}$	_____	_____
x 10	2.$\overline{857142}$	_____	_____
x 10	8.$\overline{571428}$	0.$\overline{571428}$	8

5.3D

17. Exactly the same algorithm applies to convert fractions to expansions of any base. Just multiply repeatedly by that base (as you did by 10) and record the integer parts. Complete the following table to expand 4/7 and $1.8\overline{6}$ into their repeating binary form:

	input a	fract (a)	int (a)		input a	fract (a)	int (a)
	$1.86\overline{66}$	$0.86\overline{66}$	1		4 / 7	4 / 7	0
x 2	$1.73\overline{33}$	$0.73\overline{33}$	1	x 2	8 / 7	1 / 7	1
x 2	$1.46\overline{66}$	___	___	x 2	2 / 7	___	___
x 2	$0.93\overline{33}$	___	___	x 2	4 / 7	___	___
x 2	$1.86\overline{66}$	___	___	x 2	___	___	___

Use the algorithm to express each fraction in repeating binary form.

18. 3/5

19. 4/9

20. 2/13

Use the algorithm to verify the following conversion from decimal into binary form:

21. $0.4_{ten} = 0.\overline{0110}_{two}$

22. $0.375_{ten} = 0.011_{two}$

The number in question 22 has a terminating form in base ten and in base two. It can be written in repeating-digit form in both base ten and base two.

$$0.375_{ten} = 0.375\overline{0}_{ten} = 0.374\overline{9}_{ten} = 0.011\overline{0}_{two} = 0.010\overline{1}_{two}$$

In general, every rational number can be represented in repeating or eventually repeating digit form regardless of the base used to express the number. However, their appearances may be quite different in both decimal and binary form.

5.4 THE SAW-TOOTH FUNCION AND BINARY NUMBERS

5.4A

The algorithm to convert into binary form requires repeated multiplication by 2, subtracting the integer part whenever it reaches 1. But this is exactly and precisely the definition of the saw-tooth function.

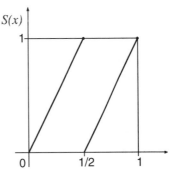

$$S(x) = \begin{cases} 2x & \text{when } 0.0 \leq x \leq 0.5 \\ 2x - 1 & \text{when } 0.5 < x \leq 1.0 \end{cases}$$

For all binary numbers $x = 0.a_1a_2..._{two}$, the first digit after the point indicates in which half of the unit interval x is located. Thus $a_1 = 0$ implies $x \leq 1/2$, whereas $a_1 = 1$ implies $x \geq 1/2$. Note that $x = 1/2$ is included in both cases, since $0.1_{two} = 0.0\overline{1}_{two}$. Using the representation $1/2 = 0.0\overline{1}_{two}$ and $1 = 0.1\overline{1}_{two}$, we obtain the following rule.

Binary evaluation of the saw-tooth function at $x = 0.a_1a_2a_3..._{two}$

For $0 \leq x \leq 1/2$, where $a_1 = 0$, move the point one place to the right.

For $1/2 < x \leq 1$, where $a_1 = 1$, move the point one place to the right and drop the unit digit before the point.

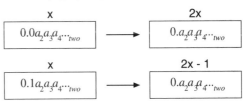

If $x_0 = 0.1_{two} = 0.0\overline{1}_{two}$, then $S(x_0) = 2x_0 = 0.\overline{1}_{two}$.
If $x_0 = 1.0_{two} = 0.\overline{1}_{two}$, then $S(x_0) = 2x_0 - 1 = 0.\overline{1}_{two}$.
If $x_0 = 0.01101_{two}$, then $S(x_0) = 2x_0 = 0.1101_{two}$.
If $x_0 = 0.1101_{two}$, then $S(x_0) = 2x_0 - 1 = 0.101_{two}$.

1. Here are the first two iterations x_1 and x_2 of the saw-tooth function with a given starting value $x_0 = 0.1011\overline{01}$ in binary form. Complete the next three iterations x_3, x_4, and x_5, and compare each iteration with x_0.

$$
\begin{aligned}
x_1 &= S(x_0) &&= 0.011\overline{01} \\
x_2 &= S^2(x_0) = S(x_1) = \underline{\qquad} \\
x_3 &= S^3(x_0) = S(x_2) = \underline{\qquad} \\
x_4 &= S^4(x_0) = S(x_3) = \underline{\qquad} \\
x_5 &= S^5(x_0) = S(x_4) = \underline{\qquad}
\end{aligned}
$$

For each of the values of x_0 expressed in binary form, find the first three iterates $x_1 = S(x_0)$, $x_2 = S(x_1) = S^2(x_0)$, and $x_3 = S(x_2) = S^3(x_0)$.

2. 0.011

3. $0.01\overline{01}$

4. $0.011\overline{011}$

5. $0.1\overline{1}$

6. $0.10\overline{10}$

7. $0.1110\overline{110}$

In our activity on kneading, we divided the dough into compartments to visualize their movement under the kneading process and to compare the results of different kneading strategies. We used subdivisions to follow the movement of grains of spice and saw how they spread through the dough. To study the corresponding iteration of points of the unit interval under the saw-tooth function, we use the more convenient binary subdivisions.

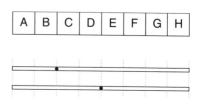

Recall that stage 3 of the binary subdivision of the unit interval contains 8 subintervals.

Third stage 0.000... 0.001... 0.010... 0.011... 0.100... 0.101... 0.110... 0.111...

0   1/2   1

Each subinterval of stage 3 corresponds to binary numbers that start with the same three digits after the point. These same three binary digits identify the interval of the stage-3 subdivision and can be used as a 3-digit label.

8. Add the missing labels to the stage-3 binary subdivision.

Third stage 0.000... 0.001... 0.010... 0.011... 0.100... 0.101... 0.110... 0.111...

001   110

Each one of the 2^n intervals of stage n of the binary subdivision can be identified by an n-digit label $a_1 a_2 a_3 ... a_n$ that corresponds to the leading n digits that all binary numbers of the interval have in common. For example, $x = 0.011010$ is contained in interval 0110 of stage 4 and in interval 01101 of stage 5 of the binary subdivision.

9. In which subinterval of stage 3 and stage 5 do you find $x = 0.110110$?

Look at the iterates of the initial value $x_0 = 0.110100\overline{110}$ under the saw-tooth function.

	Stage-3 interval	Stage-4 interval
$x_1 = S(x_0) = 0.10100\overline{110}$	101	1010
$x_2 = S(x_1) = $ _____	___	___
$x_3 = S(x_2) = $ _____	___	___
$x_4 = S(x_3) = $ _____	___	___

10. Fill in the missing values for the iterates of x_0 shown above.

11. The initial value x_0 is contained in the stage-3 interval 110 and the stage-4 interval 1101. Fill in the labels of the intervals that contain x_2, x_3, and x_4.

5.5 THE SAW-TOOTH FUNCTION AND CHAOS   5.5A

The saw-tooth function will be our role model for chaos. There are many popular scientific explanations of chaos. Some say chaos is when a process is unpredictable, when it shows sensitivity, or when errors amplify exponentially. The truth about chaos seems to be delicate. In mathematics, chaos has been characterized by the three interwoven properties of *mixing, sensitivity* and *periodicity*. We saw some glimpses of these concepts through the kneading of dough by stretch-and-fold and stretch-and-cut- and-paste and the corresponding graphical iterations of the tent and saw-tooth functions. Using binary arithmetic and the binary subdivisions, we will now be able to pin things down more completely.

MIXING — spreading the spices

One reason for kneading a piece of dough is to spread small pockets of spices throughout the dough. Similarly, we can spread the points of a small subinterval of the binary subdivision over the whole unit interval. More pre- cisely, from any given subinterval we can reach any other interval when iterating under the saw-tooth function S.

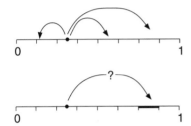

Look again at the binary subdivision of stage 3. Let us try to find an initial point x_0 in the interval 100 that reaches interval 110 after three iterations of S.

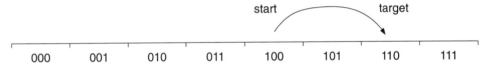

By appending the binary label of the target interval to that of the starting interval, we form $x_0 = 0.100110$. Clearly, the binary number x_0 is in the starting interval 100.

1. Find the stage-3 interval for each of the first three iterations of $x_0 = 0.100110$ under S. $x_0 = 0.100110$ is in interval 100.
    $x_1 = S(x_0) = $ _____ is in interval _____.
    $x_2 = S(x_1) = S^2(x_0) = $ _____ is in interval _____.
    $x_3 = S(x_2) = S^2(x_1) = S^3(x_0) = $ _____ is in interval _____.

2. Find the stage-3 interval for each of the first six iterations of $x_0 = 0.010100111$.

3. Find an initial point that starts in stage-3 interval 111 and reaches interval 010 after three iterations under S. Find four initial points that start in interval 010 and reach interval 111 after five iterations.

A subinterval of stage 4 corresponds to binary numbers that start with the same 4 digits after the point. Four consecutive digits in the binary expansion of an initial point identify the stage-4 intervals of corresponding iterates under S.

4. At what iteration does the initial point $x_0 = 0.111111001100111...$ first reach subinterval 0011 under S.

5. Find two initial points that start in stage-4 interval 0101 and reach interval 1100 after five iterations of S.

In general, we can find in any subinterval of $[0, 1]$ an initial point x_0 that reaches any prescribed target subinterval of $[0, 1]$ when iterated by the saw-tooth function S. More formally, the subintervals of the binary subdivision of stage n can be identified by binary labels of the form $a_1 a_2 a_3 ... a_n$. To reach subinterval $t_1 t_2 t_3 ... t_n$, from subinterval $s_1 s_2 s_3 ... s_n$ start with $x_0 = 0.s_1 s_2 s_3 ... s_n t_1 t_2 t_3 ... t_n$. In n steps through S, you reach $x_n = 0.t_1 t_2 t_3 ... t_n$.

6. Let $x_0 = 0.b_1 b_2 b_3 b_4 b_5 ... b_n c_1 c_2 c_3 c_4 c_5 ... c_n$. Find the subintervals corresponding to x_0 and the iterated points x_2, x_5 and x_n.

	Stage-2 Interval	Stage-5 Interval	Stage-n Interval
x_0 $x_2 = S^2(x_0)$ $x_5 = S^5(x_0)$ $x_n = S^n(x_0)$			

We can even determine initial points that visit a prescribed selection of subintervals or all intervals of a subdivision. Let $x_0 = 0.100110101001010000$.

7. Mark in each stage-3 subinterval the first iterate that reaches that location.

000	001	010	011	100	101	110	111
	x_1			x_0			

8. Add some digits to x_0 to find a slightly different initial point that reaches the one stage-3 interval not reached by x_0. At what iteration is it now reached under S?

9. How many stage-4 subintervals can be reached from the initial point x_0?

PERIODICITY — periodic points are dense

Not all points of a given subinterval are mixed throughout the unit interval. Rather, some of them visit a very limited selection of points and then return exactly to the starting point. This is the property of periodicity.

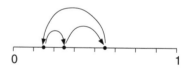

As an example, iterate the saw-tooth function starting with $x_0 = 4/7$.

| x_0 | $\to$ | x_1 | $\to$ | x_2 | $\to$ | x_3 | $\to$ | x_4 | $\to$ | x_5 | $\to$ | x_6 | $\to$ | $\cdots$ |
| 4/7 | | 1/7 | | 2/7 | | 4/7 | | 1/7 | | 2/7 | | 4/7 | | |

The value 4/7 repeats every 3 steps such that

$$x_0 = x_3 = x_6 = x_9 = ...$$

We say that x_0 is a *periodic point of period 3*. Obviously, we also have

$$x_0 = x_6 = x_{12} = ... \text{ and } x_0 = x_9 = x_{18} = ...$$

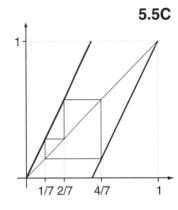

5.5C

Thus, x_0 is also a point of period 6, 9, and so on. Since x_0 first repeats itself as x_3, we say that the *minimal* period is 3. Note also that if we start the iteration with x_1 or x_2, we return after exactly three iterations with x_1 or x_2. Collectively, x_0, x_1, and x_2 are called a cycle of period 3. The path of graphical iteration nicely visualizes this fact. It is a closed path.

Iterate the following initial values and determine each minimal period.

10. $x_0 = 1/3$

11. $x_0 = 0.2_{ten}$

12. $x_0 = 2/9$

13. $x_0 = 0.\overline{011}_{two}$

14. $x_0 = 0.0\overline{011}_{two}$

15. $x_0 = 0.\overline{11010}_{two}$

Any number $x_0 = 0.\overline{a_1a_2a_3...a_n}$ between 0 and 1 with a binary form that immediately repeats digits is a periodic point of period n. Its repeating behavior under the iteration of the saw-tooth function is already visible in the repetition of the binary digits.

16. Iterate $x_0 = 0.\overline{a_1a_2a_3a_4}$ for four steps. What is its period?

$x_1 = 0.\overline{a_2a_3a_4a_1}$ $x_2 = 0.$_____ $x_3 = 0.$_____ $x_4 = 0.$_____

Where in $[0, 1]$ can we find periodic points? To answer this question, we use binary subdivisions. To find a periodic point in subinterval 101 of stage 3, take $x_0 = 0.\overline{101}$.

17. Find a periodic point in subintervals 011, 001 and 110 of the stage-3 subdivision.

18. Find a periodic point in subinterval 1101 of stage 4. List the subintervals of stage 5 that it reaches.

19. Find a periodic point that reaches every stage-3 subinterval of the unit interval.

In general, any subinterval of stage n contains periodic points under the saw-tooth function. For example, interval $a_1a_2a_3...a_n$ contains the periodic point $0.\overline{a_1a_2a_3...a_n}$. We say *periodic points are dense in the unit interval* since the size of the subintervals $(1/2)^n$ can become arbitrarily small. Thus, we can find periodic points in all $2^{10} = 1024$ subintervals of stage 10. This provides, for any point of the unit interval, a periodic point that is not more than 1/1024 apart from it. In other words, we can find a periodic point arbitrarily close to any point of the unit interval.

Numbers with repeating binary expansion are rational numbers. Indeed, all rational numbers lead to fixed points or periodic cycles for the saw-tooth function, even though some periods can be very large. However, irrational numbers have non-repeating binary expansions. Thus, they do not lead to periodic cycles when iterated.

Consider the subintervals of successive stages of the binary subdivisions.

stage	1	2	3	$\cdots$
labels	0, 1	00, 01, 10, 11	000, 001, 010, 011, 100, 101, 110, 111	$\cdots$

5.5D

Form the following irrational number by successively appending the labels:

$$x_0 = 0.0\ 1\ 00\ 01\ 10\ 11\ 000\ 001\ 010\ 011\ 100\ 101\ 110\ 111\ 0000\ 0001...$$

The iteration of this number visits all binary subintervals of all stages of the binary subdivision. For example, after 8 iterations, interval 11 of stage 2 is reached. After 8 more iteration interval 010 of stage 3 is reached. This behavior is called *ergodic*.

20. Not all irrational numbers show ergodic behavior. Which blocks of digits in x_0 can be deleted to arrive at an irrational number that will never visit subinterval 111?

SENSITIVITY — stretching apart

Even the smallest difference between initial values may lead, under iteration of the saw-tooth function, to a totally different behavior. More precisely, we will show that, arbitrarily close to a given initial point, we can find another point such that under iteration the two points will become 1/2 unit away from each other.

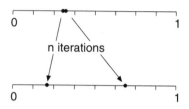

21. Recall that the length of a subinterval of the binary subdivision of the unit interval at stage n is $(1/2)^n$. At what stage will the width first be less than 1/50?

22. At what stage will the distance between two points in the same subinterval be no more than 0.00000001_{two}?

23. The point $x_0 = 0.1011\overline{010}_{two}$ is contained in subinterval 101 of stage 3 of the binary subdivision. What are the smallest and largest points of that interval? What is the interval size? What subinterval of stage 5 contains x_0? What is its size?

The distance between 0.0101 and 0.0111 is $0.001 = (1/2)^3$. Consider $x_0 = 0.1010110$ and $z_0 = 0.1010010$. What is the distance between these pairs of points?

24. x_0 and z_0 25. $S(x_0)$ and $S(z_0)$ 26. $S^2(x_0)$ and $S^2(z_0)$

Find a point that is exactly 1/2 unit away from the binary point given.

27. $0.01\overline{10}$ 28. $0.1\overline{001}$

29. The points $x_0 = 0.0110\overline{10}$ and $z_0 = 0.0111\overline{10}$ are both in interval 011 of stage 3. They are at most $(1/2)^3$ apart. What is the distance between $S^3(x_0)$ and $S^3(z_0)$?

In general, we are looking for a point z_0 that is at most $(1/2)^n$ apart from a given point $x_0 = 0.a_1a_2a_3...a_na_{n+1}a_{n+2}...$. Flip the digit a_{n+1} to a^*_{n+1}, changing it from 0 to 1 or 1 to 0, to find $z_0 = 0.a_1a_2a_3...a_na^*_{n+1}a_{n+2}...$. Both points are in interval $a_1a_2a_3...a_n$ of stage n and $(1/2)^n$ apart from each other. After n iterations we find $x_n = T^n(x_0) = 0.a_{n+1}a_{n+2}...$ exactly 1/2 unit away from $z_n = T^n(z_0) = 0.a^*_{n+1}a_{n+2}...$.

30. Find x_0 and z_0 in interval 1011 of stage 4 of the binary subdivision such that $S^4(x_0)$ and $S^4(z_0)$ are 1/2 unit apart.

5.6 THE SAW-TOOTH AND TENT FUNCTIONS   5.6A

The tent transformation requires evaluating both $2x$ and $2 - 2x$.

$$T(x) = \begin{cases} 2x & \text{when } 0.0 \le x < 0.5 \\ 2 - 2x & \text{when } 0.5 \le x \le 1.0 \end{cases}$$

Computing $2 - 2x$ is a bit more complicated than evaluating $2x$ and requires the use of duals.

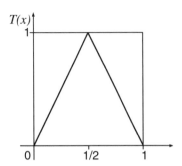

Two decimal numbers that add up to 10_{ten} are called duals with respect to base 10. For example, $x = 3.1247_{ten}$ and $x^* = 6.8753_{ten}$ are duals. Recall that 10.00_{ten} can also be written $9.9\overline{9}_{ten}$. An easy way to form the dual of a number $x = a_1.a_2a_3..._{ten}$ is to choose $x^* = a_1^*.a_2^*a_3^*..._{ten}$, where $a_k^* = 9 - a_k$. In other words, replace every decimal digit by its complement to 9.

Duals: 4.379563_{ten} and 5.620436_{ten} Check: $4.379563_{ten} + 5.620436_{ten} = 9.9\overline{9}_{ten}$

Find the dual numbers with respect to base ten.

1. $7.345\overline{73}$ 2. $0.3\overline{3}$ 3. $0.\overline{123456}$

In the same spirit, two binary numbers that add up to $2_{ten} = 10_{two} = 1.1\overline{1}_{two}$ are called duals with respect to base two. For a number $x = a_1.a_2a_3..._{two}$ between 0 and 10_{two}, find the dual by choosing $x^* = a_1^*.a_2^*a_3 *\cdot..._{two}$, where $a_k^* = 1 - a_k$. In other words, simply replace every digit 0 by 1 and every digit 1 by 0.

Find the dual of each number with respect to base two. Note that for a number in terminating digit form, first rewrite the number with repeating digits.

4. $1.010\overline{01}$ 5. $0.101\overline{100}$ 6. 1.1011

It is now easy to evaluate the term $2 - 2x$ of the tent function for $1/2 \le x \le 1$. First, compute $y = 2x$, and then take the dual $y^* = 2 - y$ with respect to base two.

$$\begin{aligned} x &= 0.1a_2a_3..._{two} \\ 2x &= 1.a_2a_3..._{two} \\ 2 - 2x &= 0.a_2^*a_3^*..._{two} \end{aligned} \qquad \begin{aligned} x &= 0.101101..._{two} \\ 2x &= 1.01101..._{two} \\ 2 - 2x &= 0.10010..._{two} \end{aligned}$$

Binary evaluation of the tent function at $x = 0.a_1a_2a_3..._{two}$

For $0 \le x < 1/2$, where $a_1 = 0$, move the point one place to the right.

For $1/2 \le x \le 1$, where $a_1 = 1$, move the point one place to the right and take the dual.

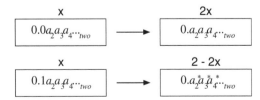

5.6B

Evaluate the tent function $T(x)$ for each value of x.

7. $0.110\overline{110}_{two}$ 8. $3/11$ 9. 0.35_{ten}

For each value of x, evaluate $T(x)$, $T^2(x)$ and $T^3(x)$.

10. $0.10\overline{100}_{two}$ 11. $0.1\overline{101}_{two}$ 12. 0.001101_{two}

Iterating the tent function requires much more care than the saw-tooth function. But there is a powerful way of reducing iterating the tent function to the much easier and transparent iteration of the saw-tooth function. Recall that we have investigated the relationship between stretch-and-cut-paste and stretch-and-fold when studying kneading. We found that two stretch-and-fold operations are equivalent to one stretch-and-cut-and-paste followed by one stretch-and-fold. This corresponds to an important relation between T and S given below:

$$T(T(x)) = T(S(x)), \text{ for } x \text{ in the interval } [0, 1]$$

13. For exploring this relationship by an example, apply the transformations to the given values for x and compare the results.

$x = 0.10101$	$x = 0.10101$	$x = 0.01101$	$x = 0.01101$
$T(x) = \underline{\qquad}$	$S(x) = \underline{\qquad}$	$T(x) = \underline{\qquad}$	$S(x) = \underline{\qquad}$
$T(T(x)) = \underline{\qquad}$	$T(S(x)) = \underline{\qquad}$	$T(T(x)) = \underline{\qquad}$	$T(S(x)) = \underline{\qquad}$

14. In Unit 4, we introduced a composition machine to determine the graph of the composition of two functions. Use this machine to determine the graph of $T(S(x))$. Be sure to evaluate the specific x values 0.0, 0.25, 0.5, 0.75 and 1.0. Compare with the graph of $T(T(x))$.

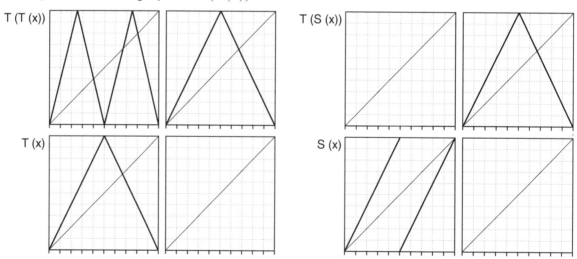

To obtain the result $T(T(x)) = T(S(x))$ in general, select an initial value $x = 0.a_1a_2a_3...$ with binary digits a_k. Then verify the equivalence for all four possible cases which depend on the choices of 0 or 1 for the values of a_1 and a_2.

15. Follow the composite transformations $T(T(x))$ and $T(S(x))$ and complete the table. Compare the final four cases for each composition.

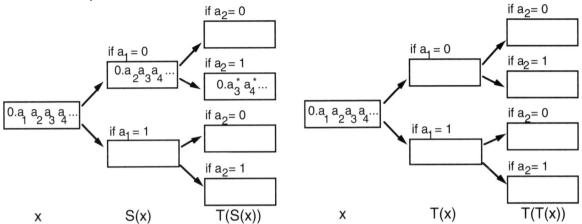

We want to reduce the complicated process of iterating the tent function T to iterating the saw-tooth function S. It turns out that n iterations of T are equivalent to $n - 1$ iterations of S followed by one iteration of T.

We can explore this fact using the graphical composition of functions as input-output machines familiar from Unit 4, that are stacked on top of each other. An input x at the bottom of a box is transformed to an output $y = f(x)$ at the top, serving as input for the next function $g(x)$. At the right, we show one stack for T, T and T and another stack for T, S and T. Using these stacks, we show graphically $T(T(T(x)))$ and $T(S(T(x)))$ for the initial value $x = 0.3$. The two final outputs are the same.

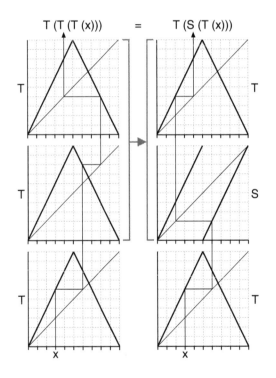

16. Draw in the two stacks the iteration of $x = 0.4$.

We have demonstrated that $T(T(x)) = T(S(x))$ for all values of x between 0 and 1. Thus, if we stack two T-icons and a T-icon on top of an S-icon, the input-output relationship of these pairs will be the same. For all input values, both pairs will produce the same output. In a stack of T and S icons, we can replace a pair of T-icons by a T-icon on top of an S-icon. Both stacks show the same input-output behavior.

17. Follow the iterations for three selected initial values of your own choice through the two stacks and compare the two results in each case.

5.6D

Now we proceed to exploit the relation $T(T(x)) = T(S(x))$ by looking at stacks with more than two transformations. As an example, take five T-icons and stack them as at the far left below. Take four S-icons, stack them, and add one T-icon on top as at the far right below. We want to compare $T(T(T(T(T(x)))))$ and $T(S(S(S(S(x)))))$ and conclude that they are the same.

18. Using the relation $T(T(x)) = T(S(x)$, two lower T-icons of the first stack have been replaced by a T-icon on top of an S-icon in the second stack. On top of these two icons draw three more T-icons. This makes the first two stacks equivalent. For any input x, the output of the two stacks will be the same.

19. Complete the third, forth and fifth stacks by replacing each marked icon pair by a pair of S- and T-icons, copying the other icons from the previous stack.

In general, we can replace a stack of n T-icons by a stack of $n - 1$ S-icons with a single remaining T-icon on top in just $n - 1$ steps. This means that n iterations of T give the same result as $n - 1$ iterations of S followed by one iteration of T.

$$T^n(x) = T(S^{n-1}(x))$$

Because each operation of S corresponds to a shift of the point, it follows that

$$T^n(0.a_1a_2...a_{n-1}a_n..._{two}) = T(0.a_na_{n+1}..._{two})$$

20. Compute $T^5(x_0)$ for $x_0 = 0.110\overline{001100}_{two}$ directly and also using the relationship between T and S. Compare the two results with each other and with x_0.

5.7 THE TENT FUNCTION AND CHAOS   5.7A

The tent function is our second model of chaos. We have seen its correspondence to the kneading of dough through the operation of stretch-and-fold. Later, this will allow us to build a bridge to the iteration of the parabola $y = 4x(1 - x)$ as introduced in Unit 4 and to nonuniform kneading operations. In this activity the relationship between the saw-tooth function and the tent function will be exploited to view once more the three basic properties of chaos. This time we will see *mixing, sensitivity,* and *periodicity* through the tent function.

Again, we use the binary numbers and the binary subdivision of the unit interval $[0, 1]$ to monitor the iterative behavior of different initial points.

MIXING — spreading the spices

As with the saw-tooth function, the tent function will take points of a small subinterval of the binary subdivision and spread them all over the whole unit interval. Likewise, there are always points from a given subinterval that move to any other subinterval when iterating the tent function T.

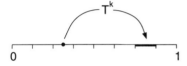

There are two cases to consider. First, let us try to get from interval 011 to interval 010 of stage 3 of the binary subdivision. We need to find an initial point x_0 in interval 011 such that $T^3(x_0)$ is in interval 010. The appropriate relationship between T and S is $T^3(x_0) = T(S^2(x_0))$. Therefore, look at two iterations of S followed by one iteration of T.

The initial marked point is $x_0 = 0.011101$.

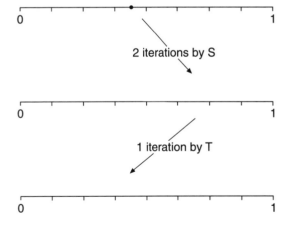

1. What subinterval of stage 3 contains x_0?

2. Mark $S^2(x_0)$. What is its value? What subinterval of stage 4 contains this point?

3. What half of the unit interval is reached by the points of subinterval 011 of stage 3 when iterated for two steps by the saw-tooth function S?

4. Mark $T(S^2(x_0))$. What is its value?

5. If z is a point of interval $1c_1c_2c_3$ of stage 4 of the binary subdivision, which interval of stage 3 contains $T(z)$?

As an example of the second case, we try to get from interval 110 to interval 010 of stage 3 of the binary subdivision. Again, use the relationship between T and S and look at two iterations of S followed by one iteration of T.

5.7B

6. Mark $x_0 = 0.110010$ as an initial point. What subinterval of stage 3 contains x_0?

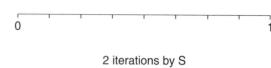

2 iterations by S

7. Mark $S^2(x_0)$. What is its value? What subinterval of stage 4 contains this point?

8. What half of the unit interval is reached by the points of subinterval 110 of stage 3 when iterated for two steps by the saw-tooth function S?

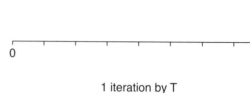

1 iteration by T

9. Mark $T(S^2(x_0))$. What is its value?

10. If z is a point of interval $0c_1c_2c_3$ of stage 4 of the binary subdivision, which interval of stage 3 contains $T(z)$?

In general, we can find, in any subinterval of $[0, 1]$, an initial point x_0 that reaches any prescribed target subinterval of $[0, 1]$ when iterated by the tent function T. More formally let x_0 be an initial value contained in interval $a_1a_2a_3...a_n$ of stage n of the binary subdivision. There are two possible cases. Take

$$x_0 = \begin{cases} 0.a_1a_2...a_nb_1b_2...b_n & \text{if } a_n = 0, \text{ or} \\ 0.a_1a_2...a_nb_1^*b_2^*...b_n^* & \text{if } a_n = 1. \end{cases}$$

Then $T^n(x_0)$ is contained in interval $b_1b_2b_3...b_n$.

11. Let $x_0 = 0.a_1a_2...a_{n-1}1b_1^*b_2^*...b_n^*$. What is $S^{n-1}(x_0)$? What is $T(S^{n-1}(x_0))$?

12. Let $x_0 = 0.a_1a_2...a_{n-1}0b_1b_2...b_n$. What is $S^{n-1}(x_0)$? What is $T(S^{n-1}(x_0))$?

13. Find x_0 in interval 1001 of stage 4 of the binary subdivision such that $T^4(x_0)$ is in interval 0110.

PERIODICITY — periodic points are dense

For the saw-tooth function, we have shown that any small subinterval of the binary subdivision contains periodic points. Thus, for all points of the unit interval, we can find a periodic point as close as we like. This means that periodic points under S are dense in the unit interval. The same periodicity property can be demonstrated for the tent function.

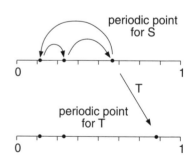

5.7C

Using the relation $T^n(x) = T(S^{n-1}(x))$, it is easy to generate periodic points for T. Let x be a periodic point for S with period n. Thus, $S^n(x) = x$.

First, apply T to the equation $S^n(x) = x$. $T(S^n(x)) = T(x)$
Next, use the relationship $T(S^n(x)) = T^{n+1}(x)$. $T^{n+1}(x) = T(x)$
Then express $T^{n+1}(x)$ as $T^n(T(x))$. $T^n(T(x)) = T(x)$

This shows that a periodic point x of the saw-tooth function provides a point $u = T(x)$ of period n for the tent function such that $T^n(u) = u$. Note that n is not necessarily the minimal period for u.

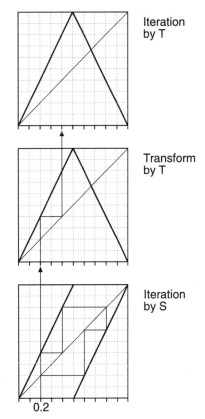

Iteration by T

Transform by T

Iteration by S

0.2

14. A periodic point for the saw-tooth function S is
    $4/7 = 0.\overline{100}_{two}$.

$$4/7 \to 1/7 \to 2/7 \to 4/7 \to \cdots$$

   A periodic point for the tent function is
   $u = T(4/7) = 6/7 = 0.\overline{110}_{two}$.

$$6/7 \to 2/7 \to 1/7 \to 6/7 \to \cdots$$

   What is the minimal period in both cases?

15. Iterate $x_0 = 0.2_{ten} = 0.\overline{0011}_{two}$ under S. The corresponding path of graphical iteration is shown in the diagram for the saw-tooth function. What is the minimal period?

16. Compute the transformed initial point $u_0 = T(u_0)$ and iterate u_0 by T for four steps. Draw the corresponding path of graphical iteration in the diagram for the tent-function shown at the top. What is the minimal period?

17. Generate a point of period 7 from $x = 0.\overline{1101110}_{two}$ for T.

How can we find a periodic point of T in a given subinterval $a_1a_2a_3...a_n$ of the binary subdivision?

We know that the point $x_0 = 0.\overline{0a_1a_2a_3...a_n}$ is a point of period $n+1$ for the saw-tooth function. That is, $S^{n+1}(x_0) = x_0$. The transformed point $u_0 = T(x_0) = 0.\overline{a_1a_2a_3...a_n0}$ is a periodic point for the tent function. But since u_0 is contained in the given subinterval, we already have found the required periodic point.

18. Find a periodic point for T in the interval 1101 of stage 4 of the binary subdivision of the unit interval.

5.7D

SENSITIVITY — stretching apart

We have seen that arbitrarily close to a given initial point
we can find another point such that, under iteration of the
saw-tooth function, the two points will become 1/2 a unit
away from each other. We now demonstrate that this is
also true for the tent function. In other words, even the
smallest difference between initial values may lead, under
iteration of the tent function, to a totally different behavior.

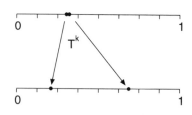

Look at the point $x_0 = 0.10011001101$. It is contained in subinterval 10011 of stage 5
of the binary subdivision. All points of this interval are at most $(1/2)^5$ apart from each
other. Now take $z_0 = 0.10011101101$ from this same interval. Using the relationship
$T^5(x) = T(S^4(x))$ we obtain:

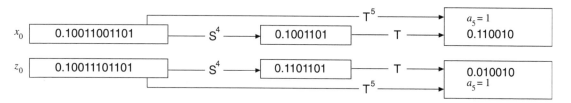

19. Find the exact distance between the two points after 5 iterations. This distance is
the difference between x_5 and z_5

In general, we are looking for a number z_0 that is at most $(1/2)^n$ apart from a given
number x_0. The points $x_0 = 0.a_1 a_2 a_3...a_n a_{n+1} a_{n+2}...$ and $z_0 = 0.a_1 a_2 a_3...a_n a_{n+1}^* a_{n+2}...$
are both contained in subinterval $a_1 a_2 a_3...a_n$ of stage n. Thus, they are at most $(1/2)^n$
apart from each other. After n iterations, $x_n = T^n(x_0) = T(S^{n-1}(x_0))$ is exactly 1/2 unit
away from $z_n = T^n(z_0) = T(S^{n-1}(z_0))$.

Indeed, after $n - 1$ iterations of S we obtain

$$S^{n-1}(x_0) = 0.a_n a_{n+1} a_{n+2}... \text{ and } S^{n-1}(z_0) = 0.a_n a_{n+1}^* a_{n+2}...$$

If $a_n = 0$, we have

$$x_n = T(S^{n-1}(x_0)) = 0.a_{n+1} a_{n+2}... \text{ and } z_n = T(S^{n-1}(z_0)) = 0.a_{n+1}^* a_{n+2}...$$

The points x_n and z_n are 1/2 unit away from each other.

20. Compute $x_n = T(S^{n-1}(x_0))$ and $z_n = T(S^{n-1}(z_0))$ if $a_n = 1$. What is the difference
between the two results?

21. Find x_0 and z_0 in interval 1011 of stage 4 of the binary subdivision such that
$x_4 = T^4(x_0)$ and $z_4 = T^4(z_0)$ are 1/2 unit apart.

5.8 POPULATION DYNAMICS   5.8A

The ultimate focus of this unit is predictability. Stable conditions are predictable and reliable. Unstable conditions are erratic and chaotic and hence unpredictable. We have seen that the dynamics of both the tent and saw-tooth functions exhibit mixing, sensitivity, and periodicity and with them, chaotic and unpredictable behavior. These same problems are associated with many scientific phenomena as well.

Population growth has become a serious world problem. Any given confined region of our globe can sustain only a certain maximal population. Accordingly, the manner in which a population subject to this maxim varies over time depends upon its own size at any given instant as well as upon the space, environmental conditions and availability of the resources needed to sustain it.

The two charts below tabulate the fraction P of the maximal population that can be sustained in each of two separate regions of the world for a period of 12 decades. Complete the plots on the grids, showing these fractions against the time t.

1.

Decade t	Fraction P of Maximum
0	0.100
1	0.208
2	0.406
3	0.695
4	0.949
5	1.007
6	0.999
7	1.000
8	1.000
9	1.000
10	1.000
11	1.000
12	1.000

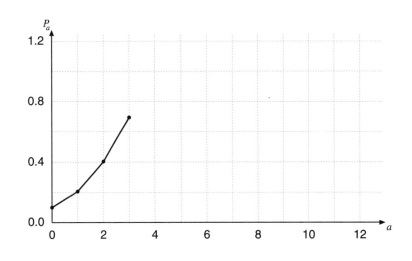

2.

Decade t	Fraction P of Maximum
0	0.100
1	0.370
2	1.069
3	0.847
4	1.236
5	0.362
6	1.054
7	0.883
8	1.193
9	0.501
10	1.252
11	0.308
12	0.948

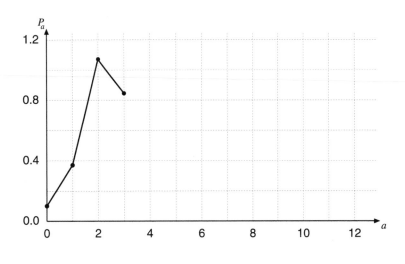

5.8B

3. Describe the long term behavior of the two populations. If these same trends for the populations continue into the future, could you predict the population sizes at the end of 100 decades? Why or why not?

4. Study carefully the behavior of the populations for various levels of P_t as given in questions 1 and 2. Under what condition will the population increase? Under what condition will it decrease?

During the mid 1800's, the Belgian mathematician Pièrre François Verhulst proposed a quadratic formula for modeling a population that changes according to the availability of resources.

$$P_{t+1} = P_t + rP_t(1 - P_t) \text{ or, equivalently, } P_{t+1} = (1 + r)P_t - rP_t^2$$

In the Verhulst model, P_t represents a fraction of the largest sustained population possible in some region. When $P_t = 0$, the population is extinct whereas if $P_t > 1$, the population has surpassed its maximum sustainable number and has to decrease again. P_{t+1} denotes the fraction during the next time period after P_t. The parameter r ranges from 0 to 3 in the model explained below.

The calculator programs below implements the Verhulst formula. The user is prompted to enter first the parameter r, and then the initial population fraction P.

Line	CASIO	Line	TEXAS INSTRUMENTS
1		1	:ClrHome
2	Fix 3	2	:Fix 3
3		3	:Disp "ENTER R"
4	"R="? → R	4	:Input R
5		5	:Disp "ENTER P"
6	"P="? → P	6	:Input P
7	0 → N	7	:0 → N
8	Lbl 1	8	:Lbl 1
9	(1+R)P−RP² → P	9	:(1+R)P−RP² → P
10	N+1 → N	10	:N+1 → N
11	N△	11	:Disp N
12	P△	12	:Disp P
13	" "	13	:" "
14		14	:Pause
15	N<12 => Goto 1	15	:If N<12
16		16	:Goto 1
17		17	:End

Lines: 1–6   set up displ* displ*, enter the parameter r and initial value P.
        7   initialize the iteration counter N to 0
       8–15   contains the iteration loop that calculates and displays the itertation count and the population size as a fraction of the maximum that can be sustained.

5.8C

Enter the appropriate program into your calculator. In the following exercises, use for initial value of P the number 0.10 corresponding to the data on sheet 5.8A.

5. Verify that the population data of question 1 assumes a parameter of $r = 1.2$ by running the program with this value assigned to r.

6. Verify that the population data in question 2 assumes a parameter of $r = 3$.

7. Use the calculator program with $r = 2.3$ to simulate the dynamics of a third population growth for 12 decades. Plot the resulting data on the grid below.

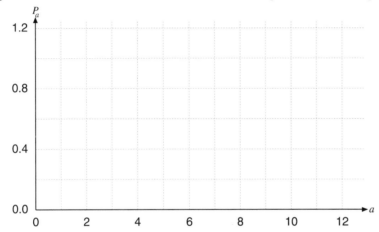

8. In what ways is the graph of question 7 different from those of questions 1 and 2? Is the long term behavior of this population predictable? What environmental conditions might cause this population to behave in this way?

The relative growth rate at time t is the change of the population fraction relative to the current size.

$$\text{relative growth at time } t: \quad \frac{P_{t+1} - P_t}{P_t}$$

Verhulst assumed that this relative growth at time t should be proportional to $1 - P_t$, the remaining fraction to reach the largest sustainable population.

$$\text{growth rate equation:} \quad \frac{P_{t+1} - P_t}{P_t} = r(1 - P_t) .$$

The factor of proportionality r can be interpreted as the fertility rate. For example, a population with a large parameter r reacts quickly to a surplus of available resources with a larger growth rate than that for a population with a smaller factor r.

When the population is small, the growth rate is positive and large. As the population approaches the 100% figure, the growth rate becomes small. When it exceeds 100%, the growth rate becomes negative and the population decreases.

9. Multiply both sides of the growth rate equation by P_t. Then add P_t. What is the result and how does it compare with the quadratic formula of Verhulst?

5.9 THE PARABOLA   5.9A

The Verhulst model for population dynamics is based on the quadratic function

$$g_r(x) = x + rx(1 - x) \,.$$

We have $P_{t+1} = g_r(P_t)$. For $r > 0$, the graph is the parabola shown here with parameter $r = 2$.

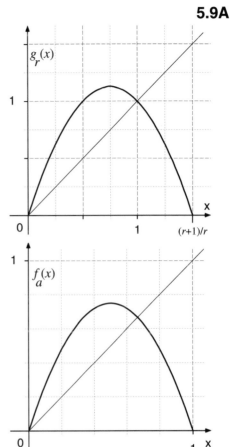

1. Verify that this parabola intersects the x-axis at $x = (r + 1)/r$. At this point, $g_r(x) = 0$.

2. Where is the intersection for $r = 2$?

In Unit 4, we studied the behavior of graphical iteration for the parabola using different parameter settings of a between 1 and 4. Shown at the right is the graph of this function for the parameter value $a = 3$.

$$f_a(x) = ax(1 - x)$$

3. Where does this parabola intersect the x-axis? Do all such parabolas with values of the parameter a between 1 and 4 intersect the x-axis at these same two points?

If we take $a = r + 1$, then the parabolas for $g_r(x)$ and $f_a(x)$ have exactly the same shape. They only differ in size. The parabola of $f_a(x)$ is a scaled down version of the one for $g_r(x)$ such that the intersection on the x-axis is at 1.

In the following exercises, use the program of Activity 5.8 to iterate the function $g_r(x)$ of Verhulst and the quadratic function $f_a(x)$ for different parameters and initial points. In order to iterate $f_a(x)$ in questions 6, 10, and 12, change line 9 of the program to

Line	CASIO	Line	TEXAS INSTRUMENTS
9	(0+R)P−RP² → P	9	:(0+R)P−RP² → P

Then the program variables P and R correspond to the value x and the parameter a.

4. Compute the next four iterates for the initial value $x_0 = 3/8$ for $g_r(x)$ and $r = 2$. For $r = 2$, the intersection of the parabola and the x-axis is at $x = (2 + 1)/2 = 3/2$ and 1/4 of that value is x_0.

    $g_2(x)$:   3/8 → _____ → _____ → _____ → _____

5. Draw the corresponding path of graphical iteration into the diagram for $g_r(x)$.

6. Compute the next four iterates of the initial value $x_0 = 1/4$ for $f_a(x)$ and $a = 3$.

    $f_3(x)$:   1/4 → _____ → _____ → _____ → _____

5.9B

7. Draw the corresponding path of graphical iteration into the diagram for $f_a(x)$.

8. Take your results for 6 and multiply by 3/2, the value of $(r+1)/r$ for $r = 2$.

    $\frac{3}{2} f_3(x) :$ 3/8 $\rightarrow$ _____ $\rightarrow$ _____ $\rightarrow$ _____ $\rightarrow$ _____

If we take $a = r + 1$, then iterating x_0 under $f_a(x)$ or iteration $(r + 1)/r$ x_0 under $g_r(x)$ are exactly equivalent. This fact will now be explored for the functions $g_3(x)$ and $f_4(x)$.

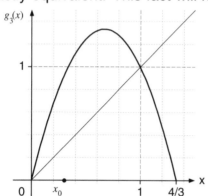   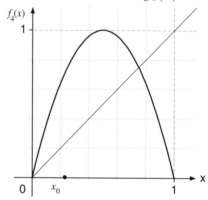

9. Compute the iterates of the initial value $x_0 = 0.3$ for $g_r(x)$ and $r = 3$. Draw the corresponding path of graphical iteration into the diagram for $g_r(x)$.

    $g_3(x) :$ 0.3 $\rightarrow$ _____ $\rightarrow$ _____ $\rightarrow$ _____ $\rightarrow$ _____

10. Compute the iterates of the initial value $x_0 = 0.225$ for $f_a(x)$ and $a = 4$. Draw the corresponding path of graphical iteration into the diagram for $f_a(x)$.

    $f_4(x) :$ 0.225 $\rightarrow$ _____ $\rightarrow$ _____ $\rightarrow$ _____ $\rightarrow$ _____

All iterates for $g_3(x)$ should be 4/3 times as large as the iterates for $f_4(x)$. Let us see what happens when we continue the iteration.

11. Compute the the following iterates of the initial value 0.3 for $g_r(x)$ and $r = 3$.

x_0	x_{10}	x_{20}	x_{30}	x_{40}	x_{50}
0.3					

12. Compute the the following iterates of the initial value 0.225 for $f_a(x)$ and $a = 4$.

x_0	x_{10}	x_{20}	x_{30}	x_{40}	x_{50}
0.225					

Theoretically, all the iterates for $g_3(x)$ should be 4/3 times as large as the iterates for $f_4(x)$. However, observe that for the last values of x_{50} the computations seem to contradict this rule. Something seems to be wrong! The answer is amazing. Sensitivity has knocked out the calculator! Small rounding errors have been amplified so dramatically that neither the higher iterates of $g_3(x)$ nor the iterates of $f_4(x)$ can be considered to be correct.

5.10 SIGNS OF CHAOS FOR THE PARABOLA   5.10A

Our current knowledge of chaotic behavior has come, in large part, through recent developments in technology. Computers help us see that a single system can have both stable, predictable behavior and chaotic, unpredictable behavior. They also help us find the points of transition between the two. When an iterative process is chaotic, it displays three key properties.

The Characteristics of Chaos

Sensitivity — Sensitive dependence upon initial conditions implies that any alteration of the initial state generates dramatically different outcomes in the iterated sequence.

Mixing — Mixing suggests that the points of any small interval eventually spread under iteration over the whole domain.

Periodic Points — The abundance of periodic points refers to the presence of infinitely many periodic points as well as infinitely many nonperiodic points distributed throughout every small interval.

The technology of the graphing calculator can be helpful in studying chaotic behavior. This iteration program displays related behavior for the quadratic function.

Line	CASIO	Line	TEXAS INSTRUMENTS
1		1	:ClrDraw
2	0 → R	2	:0 → R
3	Range 0, 1, 1, 0, 1, 1	3	:0 → Xmin
4		4	:1 → Xmax
5		5	:0 → Ymin
6		6	:1 → Ymax
7	"A="? → A	7	:Disp "A="
8		8	:Input A
9	"I="? → I	9	:Disp "I="
10		10	:Input I
11	Graph Y=AX(1−X)	11	:DrawF AX(1−X)
12	Graph Y=X	12	:DrawF X
13		13	:AI-AII → J
14	Plot I,0	14	:Line(I, 0, I, J)
15		15	:Goto 2
16	Lbl 1	16	:Lbl 1
17	AI−AII → J	17	:AI−AII → J
18	Plot I,J	18	:Line(I, I, I, J)
19		19	:Lbl 2
20	Plot J,J	20	:Line(I, J, J, J)
21	Line △	21	:Pause
22	J △	22	:Disp J
23		23	:Pause
24	R+1 → R	24	:R+1 → R
25	J → I	25	:J → I
26	R<8=>Goto 1	26	:If R<8
27		27	:Goto 1
28		28	:End

5.10B

1. Compute the iterations of $x_0 = 0.08$ and $x_0 = 0.10$ to the first three digits.

   $0.08 \rightarrow$ _____ $\rightarrow$ _____ $\rightarrow$ _____ $\rightarrow$ _____ $\rightarrow$ _____

   $0.10 \rightarrow$ _____ $\rightarrow$ _____ $\rightarrow$ _____ $\rightarrow$ _____ $\rightarrow$ _____

2. Which of the properties of chaos does this behavior represent?

The domain intervals I represent intervals from which initial points are chosen for iteration through $f(x) = 4x(1 - x)$. For each domain interval I, determine the complete range interval R that would be generated by subjecting I to one step of graphical iteration through f. Shade the path from I to R on each graph as shown.

Iteration 0, $I = [0.08, 0.10]$

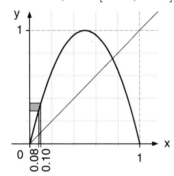

3. Iteration 1, $I = [0.29, 0.36]$

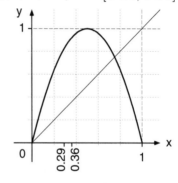

4. Iteration 2, $I = [0.83, 0.92]$

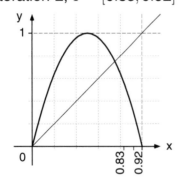

5. Iteration 3, $I = [0.56, 0.29]$

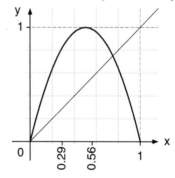

6. Iteration 4, $I = [0.82, 1.00]$

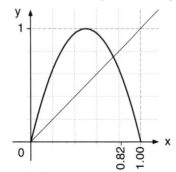

7. Iteration 5, $I = [0.00, 0.59]$

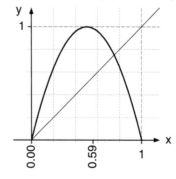

5.10C

The iteration stages defined in the example and questions 3–7 correspond to successive applications of graphical iteration, starting from initial points within the small interval $[0.08, 0.10]$ of stage 0.

8. By iteration 5, the points of the initially small interval become spread over the whole domain. Which of the properties of chaos does this behavior represent?

9. The graphs below display the iteration path for four distinct initial points p_0, q_0, r_0, and s_0. Determine from the graph the period of each of these four initial points.

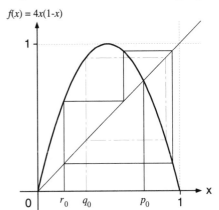
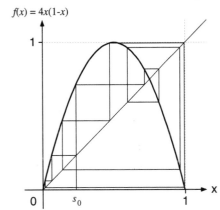

We can algebraically locate periodic points within the domain of $f(x) = 4x(1 - x)$ by utilizing the fact that applying function f to periodic x_0 finitely many times must result in the reappearance of x_0.

10. If an initial point is a fixed point, then $x_0 = f(x_0)$. In this case, $x_0 = 4x_0(1 - x_0)$. Solve this equation and check that the resulting two points are indeed fixed points by substituting their values into $f(x)$ and simplifying.

11. Explain why fixed points of f will also satisfy the equation $x_0 = f(f(x_0))$.

12. In order to find initial points having period 2, we must solve the equation $x_0 = f(f(x_0))$ for x_0. Complete the process of solving this equation initiated below. Use the fact that fixed points obtained in question 10 also satisfy $x_0 = f(f(x_0))$.

$$f(f(x_0)) = x_0$$
$$f(4x_0(1 - x_0)) = x_0$$
$$4(4x_0(1 - x_0))(1 - 4x_0(1 - x_0)) = x_0$$
$$-64x_0^4 + 128x_0^3 - 80x_0^2 + 15x_0 = 0$$

13. What equations would we solve to locate points having period 3, period 4, and period n?

Although these equations would be difficult to solve, they suggest the presence of infinitely many periodic initial points. But does this mean that we can find periodic points in any small subinterval in $[0, 1]$? We will answer this question in the next activity.

5.11 TRANSFORMATION FROM TENT TO PARABOLA   5.11A

We have seen some examples indicating that the iteration of the quadratic function $f(x) = 4x(1 - x)$ appears to have those same characteristics of chaos as the tent and saw-tooth functions. If we could be certain that the quadratic function exhibits sensitivity, mixing and dense periodic points, then the ability to predict its long term behavior would become virtually impossible. Since many iterative phenomena are best modeled by functions such as the quadratic, it is probable that we encounter the same problems in predicting the long term behavior of these phenomena. Many people believe such is the case with most natural systems. The unreliability of weather prediction is just one familiar example.

The tent function is a mathematical model for kneading by stretch-and-fold. If we uniformly subdivide the unit interval into 8 subintervals, the transformed images are all exactly twice as large as the original and their positions are the same as if we had applied the stretch-and-fold operation. We now transform the same intervals by the parabola.

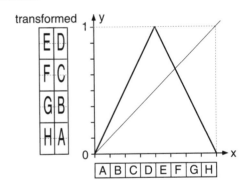   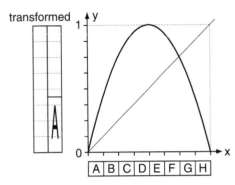

1. Draw the transformed images for the parabola and complete the labels A – H.

2. Which transformed intervals are the smallest? Which are the largest?

The transformation by the parabola stretches the outer subintervals and compresses the inner ones. We can try to compensate for this effect by using a non-uniform subdivision. The figure shows the same subdivision into 8 subintervals along the x-axis and the y-axis.

3. Draw and label the images of the intervals A – H for this non-uniform subdivision on the x-axis.

4. How many parts of the y-axis subdivision does the transformed image of B cover. Do all images cover the same number of parts?

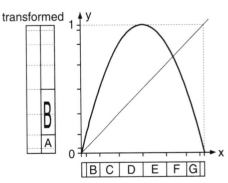

The non-uniform subdivison used in exercises 3 and 4 is determined by the trigonometric function

$$h(x) = \sin^2\left(\frac{\pi x}{2}\right) .$$

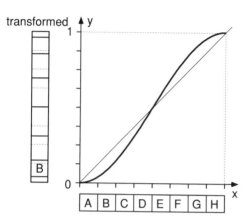

Part of its graph is shown in the figure. This function is a one-to-one map for the points of the unit interval. It transforms a uniform subdivision into a subdivision of the form used in exercises 3 and 4.

5. Draw the transformed images of the uniform subintervals A – H. Compare them with the subintervals at the x-axis of exercise 3.

Let us now use the function $h(x)$ to transform points of the unit square. We map each point (x, y) to point $(x', y') = (h(x), h(y))$. The left figure shows a uniform grid. The vertical and horizontal lines are given for x- and y-values of 1/8, 2/8, 3/8, ..., 7/8.

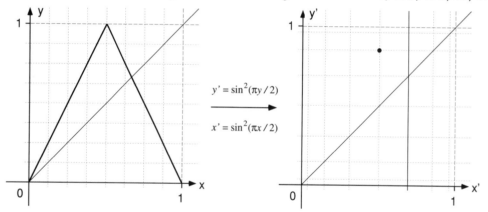

6. The right figure shows the transformed image of the same grid. The image of the vertical line $x = 5/8$ is drawn in solid. Mark in the same fashion the images of the vertical line $x = 2/8$ and the horizontal line $y = 3/8$.

7. We also have marked the point $(h(0.5), h(0.75))$. Mark $(h(0.125), h(0.75))$ and $(h(0.75), h(0.25))$.

8. The figure on the right again shows the transformed grid. Draw onto this grid the transformed points $(h(x), h(y))$ of all intersection points (x, y) of the tent function with the vertical lines of the original uniform grid. Connect the points.

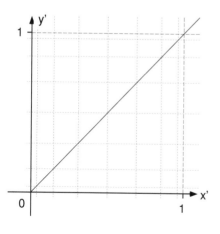

5.11C

The connected points form a rough approximation of the parabola $y = 4x(1 - x)$. If we were to use a finer grid we would obtain a better approximation. Indeed, if we take all the points (x, y) of the graph of the tent function $y = T(x)$, the transformed images $(x', y') = (h(x), h(y))$ form the graph of the parabola $y' = 4x'(1 - x')$.

The left figure shows the tent function on a uniform grid with spacing in tenths and some steps of graphical iteration for $x_0 = 0.3$. The right figure shows the transformed image of this grid and the parabola $y = 4x(1 - x)$.

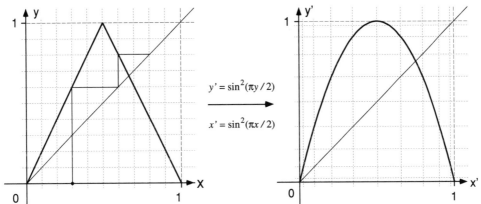

9. Draw the transformed image of the path of graphical iteration.

Observe that the transformed path of exercise 9 is exactly that path obtained from graphical iteration of $\sin^2(0.3\pi/2) = 0.220245...$ on the parabola. The transformed image of the path of graphical iteration for any x_0 under the tent function is always the path of graphical iteration for the parabola for the corresponding initial value x_0'. Use your calculator to compute the values of the sine function and the parabola needed below.

10. Consider $x_0 = 0.4$ for $T(x)$. Find the corresponding value x_0' for the parabola.

11. We know that $x_0 = 2/7$ generates a period-3 cycle $x_0 \to x_1 \to x_2 \to x_3$ under the tent function. Find the transformed values $x_k' = h(x_k)$ for all points. Compute the iteration of x_0' under the parabola $f(x)$. Compare your results.

$$x_0 = 2/7 \qquad x_0' = h(x_0) = \underline{\hspace{2cm}} \qquad\qquad x_0' = \underline{\hspace{2cm}}$$
$$x_1 = T(x_0) = 4/7 \qquad x_1' = h(x_1) = \underline{\hspace{2cm}} \qquad\qquad f(x_0') = \underline{\hspace{2cm}}$$
$$x_2 = T^2(x_0) = 6/7 \qquad x_2' = h(x_2) = \underline{\hspace{2cm}} \qquad\qquad f^2(x_0') = \underline{\hspace{2cm}}$$
$$x_3 = T^3(x_0) = 2/7 \qquad x_3' = h(x_3) = \underline{\hspace{2cm}} \qquad\qquad f^3(x_0') = \underline{\hspace{2cm}}$$

12. The initial point $x_0 = 8/25$ generates a period-10 cycle $x_0 \to x_1 \to \cdots \to x_{10}$ when iterated under the tent function as shown by the values in the table on the next page. Find the transformed values $x_k' = h(x_k)$ for all points. Use all digits of x_0' for its iteration under the parabola $f(x)$ and record the iterates to 3 digits. Compare your results.

k	x_k	$x'_k = \sin^2(x_k \frac{\pi}{2})$	$f^k(x'_0)$		5	6/25		
0	8/25				6	12/25		
1	16/25				7	24/25		
2	18/25				8	2/25		
3	14/25				9	4/25		
4	22/25				10	8/25		

Iterating an inital point x_0 under the tent function and iterating the transformed point $x'_0 = \sin^2(x_0\pi/2)$ under the parabola $f(x) = 4x(1-x)$ produces iterations that correspond to each other by means of the transformation $x' = h(x) = \sin^2(x\pi/2)$.

To establish this algebraically, start with x_0 for the tent function and use $y_0 = x'_0$ for the parabola. Thus, $x_0, x_1, \ldots$ is the iteration under the tent function and $y_0, y_1, \ldots$ is the corresponding iteration under the parabola. We can show by induction that, in fact, $y_k = x'_k = h(x_k)$ for *all* numbers $k = 0, 1, 2, \ldots$, proving the equivalence.

13. a. Start with the transformation $y_0 = \sin^2(x_0\pi/2)$, where $0 \leq x_0 \leq 1$. Substitute for y_0 in the formula for the quadratic iteration, $y_1 = 4y_0(1 - y_0)$.

   $y_1 = $ _____

   b. Substitute using the trigonometric identity $\cos^2\alpha = 1 - \sin^2\alpha$.

   $y_1 = $ _____

   c. Simplify using the double-angle identity $\sin 2\alpha = 2\sin\alpha\cos\alpha$.

   $y_1 = $ _____

   d. Show that $x_1 = T(x_0)$ satisfies the equation $y_1 = \sin^2(x_1\pi/2)$, for the y_1 found above. Start with the case $0 \leq x_0 \leq 1/2$. Thus, $x_1 = T(x_0) = 2x_0$.

   $\sin^2(x_1\pi/2) = $ _____

   e. Now do the other case, $1/2 < x_0 \leq 1$. First substitute $x_1 = T(x_0) = 2 - 2x_0$.

   $\sin^2(x_1\pi/2) = \sin^2($ _____ $) = \sin^2($ _____ $)$

   f. Simplify, first using $\sin^2(\alpha + \pi) = \sin^2(\alpha)$, and then $\sin^2(-\alpha) = \sin^2\alpha$.

   $\sin^2(x_1\pi/2) = \sin^2($ _____ $) = \sin^2($ _____ $)$

The result shows $y_1 = x'_1$ and the conclusion now follows by induction. Thus, the iteration of the tent function and the parabola are totally equivalent. All the signs of chaos are found when iterating $f(x) = 4x(1 - x)$.

- Points that exhibit sensitivity for the tent function transform into points that show sensitivity for the parabola.
- Points that show mixing by leading from one given interval to another for the tent function transform into points that have the same behavior for the parabola.
- Points that are periodic for the tent function transform into points that are periodic for the parabola.

5.12 LONG TERM BEHAVIOR AND TIME SERIES

5.12A

Many iterative processes only exhibit their essential behavior after an extensive sequence of iterations. This activity develops a new format for viewing such situations.

The dynamics of population change can be viewed directly through graphical iteration on the appropriate quadratic models. These two use parameter values of 1.8 and 4.

Graphical iteration for
$f(x) = 1.8x(1 - x)$ when $x_0 = 0.2$

Graphical iteration for
$f(x) = 4x(1 - x)$ when $x_0 = 0.2$

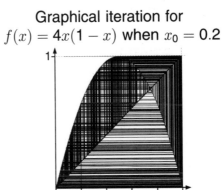

1. Describe the behavior through fifty successive iterations as shown in the two graphs above. Would you expect to learn much more from additional iterations?

Our major interest here is to discover patterns that only become apparent in the long run. In order to better study functions that model this phenomena through long term behavior, we plot successive iterates against their corresponding iteration counts. We show here the same data as above but this time viewed differently by successively plotting the time series $(0, x_0)$, $(1, x_1)$, $(2, x_2)$, $(3, x_3)$, $(4, x_4)$, ... , (n, x_n).

Time series for
$f(x) = 1.8x(1 - x)$ when $x_0 = 0.2$

Time series for
$f(x) = 4x(1 - x)$ when $x_0 = 0.2$

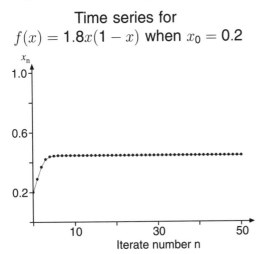

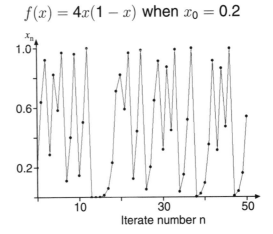

2. Compare the long term behavior for $f(x) = 1.8x(1 - x)$ graphed above on the left to that of $f(x) = 4x(1 - x)$ graphed above on the right. Which one appears to have a stable and predictable long term behavior? Which one appears to have a chaotic and unpredictable long term behavior?

5.12B

The setting of parameter a in $f(x) = ax(1 - x)$ has a profound influence upon the long term iteration behavior of the function. The calculator programs below plot the time series for functions of this form.

Line	CASIO	Line	TEXAS INSTRUMENTS
1		1	:ClrDraw
2	Range 0,50,5,0,1,0.1	2	:0 → Xmin
3		3	:50 → Xmax
4		4	:5 → Xscl
5		5	:0 → Ymin
6		6	:1 → Ymax
7		7	:0.1 → Yscl
8	"COEF A"? → A	8	:Disp "COEF A"
9		9	:Input A
10	"INIT PT"? → I	10	:Disp "INIT PT"
11	Plot 0,I	11	:Input I
12	0 → C	12	:0 → C
13	Lbl 1	13	:Lbl 1
14	AI(1−I) → J	14	:AI(1−I) → J
15	Plot C+1,J	15	:Line(C,I,C+1,J)
16	J → I	16	:J → I
17	Isz C	17	:IS>(C,50)
18	C<50=>Goto 1	18	:Goto 1
19	Line △	19	:End

Execute the program on your graphing calculator for each of the following values of the parameter a. Describe verbally the long term behavior of the iteration after using a variety of different initial values.

3. $a = 1.90$ 4. $a = 2.75$ 5. $a = 3.25$

6. Has any significant change in long behavior taken place as the parameter a increased from 1.90 to 3.25? Does the choice of an initial point appear critical?

7. $a = 3.50$ 8. $a = 3.75$ 9. $a = 4.00$

10. Summarize the effect upon long term behavior as the parameter a in $f(x) = ax(1 - x)$ increases from $a = 1.90$ to $a = 4.00$. For what values of the parameter a can the most accurate prediction of long term behavior be made?

The long term behavior of phenomena modeled by functions of the sort studied in this activity generally does not depend upon the initial point. Rather, it depends upon the parameter that establishes the features of such phenomena. The change of the parameter can lead to significantly different behavior.

For example, the main issue in population dynamics is not the initial size of the population but the conditions that establish the growth rate for the population. Different settings of parameters may lead to a predictable development or erratic, unpredictable fluctuations in the population.

5.13 THE FEIGENBAUM PLOT   5.13A

We have seen from the last activity that the long term iteration behavior of the quadratic function $f(x) = ax(1 - x)$ can vary dramatically as the parameter a varies. Depending upon the value assigned to parameter a, the long term behavior of $f(x)$ can show anything from substantially stable to erratic outcomes. This activity focuses on how and where the changes in behavior occur as a increases from 1 through 4.

The following plots show the different behaviors of $f(x) = ax(1 - x)$ for three different values of the parameter a. In each case, 50 iterates are shown starting at $x_0 = 0.2$.

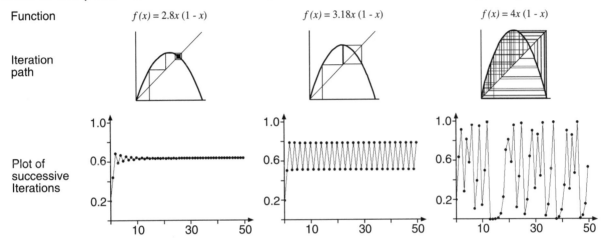

1. Use pencil and ruler to perform four steps of graphical iteration on the graph of $f(x) = 3.5x(1 - x)$ shown below at the left, starting from the initial point $x_0 = 0.125$.

2. Check your work in question 1 using the graphing calculator program given in Activity 5.12. When prompted in the program, set the increment to 0.

3. On the axes below at the right, plot the long term behavior of $f(x) = 3.5x(1 - x)$ as done in the examples above. Use the graphing calculator to generate the values.

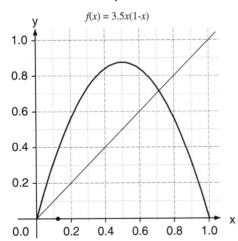

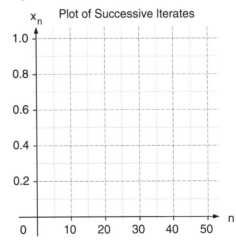

5.13B

Each setting of parameter a illustrated thus far leads to a different long term behavior. For $a = 2.8$, the sequence of iterates stabilizes on $x_n = 0.64...$ as n increases. Use the graphs from the previous page and your calculator to write the corresponding descriptions of the long term behaviors for these values of the parameter a.

4. $a = 3.18$

5. $a = 3.5$

6. $a = 4.0$

The set of values that repeatedly appear and upon which the sequence of iterates eventually settles is called an *attractor*. Some parameter values such as $a = 2.8$ have a fixed point attractor associated with them. For other settings of parameter a, such as $a = 3.18$, the iterates oscillate back and forth between two points which together form an attractive 2-cycle. Still other parameter values yield long term behaviors that move periodically among more points or even continue to erratically fill an entire interval. Describe the attractors for the following values of parameter a.

7. $a = 2.8$ 8. $a = 3.18$ 9. $a = 3.5$ 10. $a = 4$

A graph can be drawn to show the values of the various attractors for different values of the parameter a. As a moves from 1 through 4 on the horizontal axis, the values for the different points in the corresponding attractors are plotted as heights above the associated values of a.

11. The fixed point attractor for $a = 2.8$ and the 2-cycle attractor for $a = 3.18$ are shown on the plot below. Plot the attractors for $a = 3.5$ and $a = 4$.

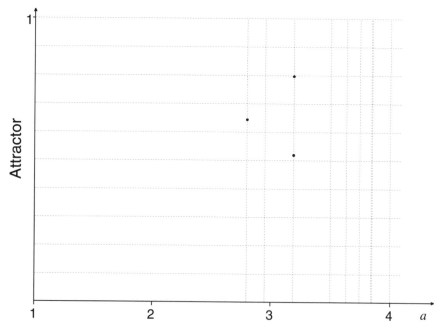

5.13C

Long term behavior is not easy to recognize during the first few stages of iteration. This program automatically iterates $f(x) = ax(1 - x)$ 100 times before it begins to display subsequent iterates.

Line	CASIO	Line	TEXAS INSTRUMENTS
1	$0 \rightarrow$ C	1	:$0 \rightarrow$ C
2	$0.2 \rightarrow$ X	2	:$0.2 \rightarrow$ X
3	"COEF A"? $\rightarrow$ A	3	:Disp "COEF A"
4		4	:Input A
5	Lbl 1	5	:Lbl 1
6	AX(1−X) $\rightarrow$ X	6	:AX(1−X) $\rightarrow$ X
7	Isz C	7	:IS>(C,100)
8	C<101=>Goto 1	8	:Goto 1
9	X △	9	:Disp X
10		10	:Pause
11	C<115=>Goto 1	11	:If C<115
12		12	:Goto 1
13		13	:End

12. Execute the appropriate program on your calculator using each of the parameter values given. Study the behavior after 100 iterations and, where possible, identify the points that form the attractor. Record the points in the table below, and then plot the attractors above the corresponding parameter values on the plot on the preceding page for question 11.

Parameter a:	2.95	3.63	3.74	3.84	3.846
Points that form the attractor					

13. Over what part of the interval $1 \leq a \leq 4$ does there appear to be a fixed point attractor? Where would you expect to find 2-cycle attractors? What about 3-cycle attractors and 4-cycle attractors?

14. At this point, the plot for question 11 should reveal some of the difficulties in predicting long term results when iterating a specific initial point through the quadratic function $f(x) = ax(1 - x)$. Explain why it becomes increasingly difficult to make these predictions as the value of the parameter a increases towards 4.

5.13D

The quadratic function $f(x) = ax(1 - x)$ clearly exhibits an extraordinary variety of changing behaviors as the parameter a increases in value from 1 through 4. The most significant connection is with predictability. A stable, predictable long term behavior exists in the form of a fixed point attractor for low values of a. As a increases, predictability becomes more complex as the number of points in the attractors increases. As a approaches 4, the behavior moves quickly into an erratic, chaotic, unpredictable state.

The consequence of this observation is profound and disturbing. Natural phenomena governed by such a quadratic relationship may appear to be in or out of control solely because of the functional relationship among the parameters and variables. Under one set of conditions they can be totally predictable yet, for only slight changes in conditions, they become completely chaotic and unpredictable.

A completed plot of attractors for all functions of the form $f(x) = ax(1 - x)$ for every parameter value from 1 through 4 is shown below. This complete plot is named after the American physicist Mitchell J. Feigenbaum. His work during the mid-1970's at Los Alamos Laboratory highlighted important properties associated with plots of this type.

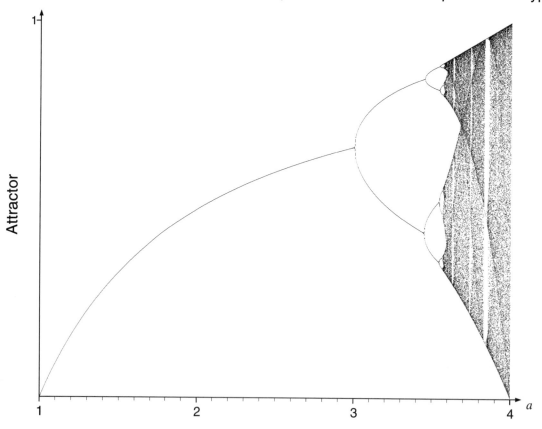

5.13E

Precisely at $a = b_1 = 3$ there is the transition of the attractor from a fixed point to a 2-cycle. This starts a whole sequence of so-called *period-doubling bifurcations*. At $a = b_2 = 3.4494...$ the 2-cycle changes into a 4-cycle, at $a = b_3 = 3.544...$ the 4-cycle becomes an 8-cycle, and so on. The upper figure on the right shows the Feigenbaum plot for the parameter range from the first bifurcation at $a = b_1 = 3$ to $a = 3.569945...$ which is approximately the point where the sequence of period-doublings ends. We call this part of the plot the period-doubling tree. The lower figure is a close up that shows the part of the period-doubling tree starting at $a = 3.4494...$ The location of the bifurcation points in both figures match almost exactly on the parameter axes. In terms of a formula this means

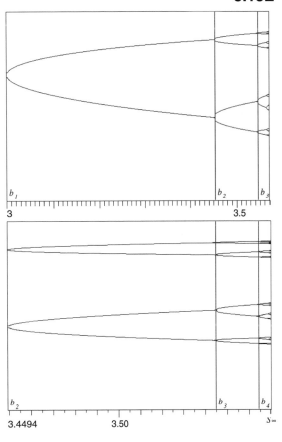

$$\frac{b_2 - b_1}{b_3 - b_2} \approx \frac{b_3 - b_2}{b_4 - b_3} \approx \frac{b_4 - b_3}{b_5 - b_4} \approx \cdots$$

Let us explore this property.

15. Here is a list of the first seven bifurcation points. Compute the differences.

$$b_1 = 3.0$$
$$b_2 = 3.449490... \qquad d_1 = b_2 - b_1 = \text{_____}$$
$$b_3 = 3.544090... \qquad d_2 = b_3 - b_2 = \text{_____}$$
$$b_4 = 3.564407... \qquad d_3 = b_4 - b_3 = \text{_____}$$
$$b_5 = 3.568759... \qquad d_4 = b_5 - b_4 = \text{_____}$$
$$b_6 = 3.569692... \qquad d_5 = b_6 - b_5 = \text{_____}$$
$$b_7 = 3.569891... \qquad d_6 = b_7 - b_6 = \text{_____}$$

16. Based on these results compute the ratios of the successive differences d_k/d_{k+1}.

d_1/d_2	d_2/d_3	d_3/d_4	d_4/d_5	d_5/d_6

5.13F

Note, that these ratios from bifurcation to bifurcation are not exactly identical but seem to tend to a certain value. In fact, this is one of the great discoveries of Feigenbaum. The ratios converge with increasing k to 4.669202..., which is called the *Feigenbaum number*.

$$\lim_{k \to \infty} \frac{d_k}{d_{k+1}} = \delta = 4.669202...$$

At first, it seems that this is just another number which documents the behavior of our particular example, the quadratic iteration. However, Feigenbaum went on to demonstrate that the character of δ is quite different. The number $\delta = 4.669202...$ is *universal*. It is the same for a wide range of different iterators. For example, it occurs in the logistic equation (see Activity 5.8) or the iteration of the trigonometric function $g_a(x) = ax^2 \sin(\pi x)$ with parameter a.

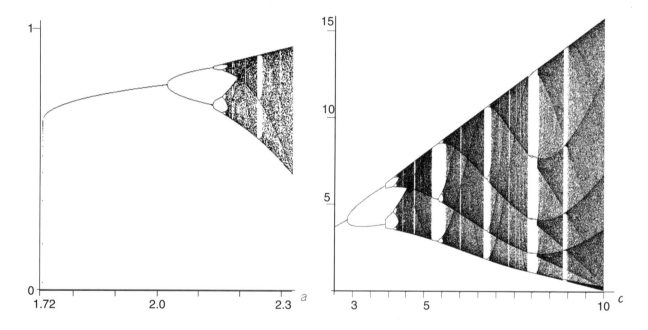

The constant δ can also be found in the study of certain differential equations and even in physical experiments ranging from turbulent flow to electronic circuits. The left figure shows the Feigenbaum plot for $g_a(x)$. It looks very similar to the plot of the quadratic iterator. The right figure shows the plot for a set of differential equations, the so called Rössler system.

In all these cases we can observe the period-doubling bifurcation and the analysis of the sequence of bifurcation points leads to the universal Feigenbaum constant δ.

Unit 6
The Mandelbrot Set

KEY OBJECTIVES, NOTIONS, and CONNECTIONS

Cartographers often construct maps that utilize colors to highlight similar regions relative to a particular feature of interest such as elevation or the amount of annual precipitation. In a similar manner, the activities in this unit focus upon the distinctive behaviors under graphical iteration of functions from a given quadratic class. The resulting colored Mandelbrot map of these behaviors exhibits surprisingly intricate and incredibly beautiful structure.

Connections to the Curriculum

The material covered in these strategic activities relates to many of the concepts found in a contemporary mathematics program. They may be presented separately or integrated into the existing curriculum through those areas to which they are connected.

PRIMARY CONNECTIONS:

Evaluating Functions	Quadratic Functions
Coordinate Geometry	Graphing
Function Composition	Visualization
Geometric Patterns	

SECONDARY CONNECTIONS:

Transformations	Conic Sections
Complex Numbers	Mappings
Graphing Calculator	Piecewise Functions
Numerical Patterns	Limit Concept

Underlying Notions

Prisoner Set

An initial point x is a prisoner if iteration of x through a given function f yields a sequence of iterates that fails to escape to infinity.

Invariant Interval

An interval $[a, b]$ is invariant and therefore a part of the prisoner set P of function f if $f(x)$ is an element of $[a, b]$ whenever x is an element of $[a, b]$.

Barrier

A value of a parameter is referred to as a barrier if it is on the boundary between two distinctly different types of behavior.

Critical Point

A point x in the domain of a quadratic function f is critical if the value of function f at x is a maximum or minimum.

Orbit

The orbit of a point x refers collectively to the entire sequence of iterates obtained from iterating x through some function.

Julia Set

For a specific value of parameter c in $f(z) = z^2 + c$, the points which lie on the boundary between the prisoner set and those points which escape to infinity are collectively referred to as the Julia set.

MATHEMATICAL BACKGROUND

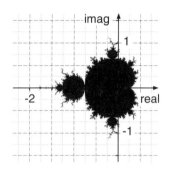

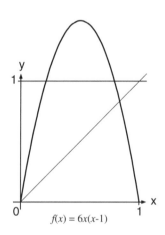

$f(x) = 6x(x\text{-}1)$

The Bigger Picture

Many applications use maps to describe the condition of a given location relative to a particular feature. A geographical map may plot temperatures across the country at a given time. In medicine, an x-ray of body tissue can locate areas of greater or lesser density and map them using darker or lighter shades of gray.

For any point (x, y) on such a map, the numerical or visual indicator describes the condition at that point relative to the feature of interest.

The set of points shaded in black in the figure to the left forms the Mandelbrot set. This map describes for each point in the complex plane the effects of a single associated function when performing iteration. When iteration generates a connected collection of prisoner points, the single point on the plane related to that given function lies within the Mandelbrot set and it is colored black.

Prisoner Sets

Some domain points under graphical iteration produce sequences of iterates that approach infinity. Other domain points for the same function yield sequences that fail to escape to infinity. These latter points are collectively referred to as a prisoner set. In this unit, we seek to characterize prisoner sets for two special types of quadratic functions according to whether the related prisoner sets form a connected region or a totally disconnected dusting of points.

When performing graphical iteration on quadratic functions of the form $f(x) = ax(1-x)$, the prisoner set can exhibit itself in one of two forms. Either the prisoner set consists of an invariant interval or the set is totally disconnected. Specifically, the prisoner set consists of:

- the invariant interval [0,1] when $1 \le a \le 4$, or

- a totally disconnected set of points when $a > 4$.

Thus, among functions of the form $f(x) = ax(1-x)$, when parameter a passes from below the barrier of $a = 4$ to above it, a transition occurs from connected to disconnected prisoner sets.

Even without reference to the value of parameter a, it is possible to determine which of these two outcomes will occur. If iteration of the critical initial point $x = 0.5$ through $f(x) = ax(1 - x)$ gives a sequence of values that escapes to infinity, then the prisoner set will be disconnected. Otherwise, the prisoner set will be the invariant interval [0,1].

From $y = ax(1 - x)$ To $y = x^2 + c$

The Mandelbrot set and its related Julia sets are all developed by reference to functions that have the form $y = x^2 + c$. Fortunately, the behaviors exhibited by points under iteration through functions of the prior form $y = ax(1 - x)$ faithfully replicate the behaviors of points under iteration through functions defined by $y = x^2 + c$. In fact, a one-to-one structure preserving mapping can be established between the two classes of quadratic functions. Thus, those features known in the class already studied readily transfer through the mapping to the other.

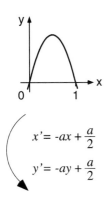

Specifically for each function defined by $y = ax(1 - x)$, a related quadratic function $y = x^2 + c$ can be obtained by a sequence of three geometric transformations. Begin with the parabola $y = ax(1 - x)$. Rotate it 180 degrees, magnify the result by a factor of a, and then translate $a/2$ units up and $a/2$ units to the right. A parabola appears that opens upwards and has the y-axis as its axis of symmetry. The foregoing sequence of geometric transformations may be accomplished algebraically by applying the two formulas shown at the left for computing image coordinates (x', y') from the given original coordinates (x, y). These equations not only map original points (x, y) to the image points (x', y'), but they also map the parabola $y = ax(1-x)$, to the image parabola $y = x^2 + (a/2 - a^2/4)$. This defines parameter c in terms of parameter a as $c = a/2 - a^2/4$.

$$x' = -ax + \frac{a}{2}$$

$$y' = -ay + \frac{a}{2}$$

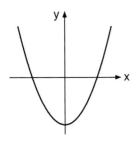

The correspondence between the two classes of quadratic functions indicates that the characterization given for prisoner sets within the $f(x) = ax(1 - x)$ class of functions also exists within the $f(x) = x^2 + c$ class. Within the $y = x^2 + c$ class of quadratic functions, prisoner sets are either connected or totally disconnected. Moreover, it follows that since $y = 4x(1 - x)$ maps to $y = x^2 - 2$ through the transformations above, the parameter $c = -2$ must act as the barrier dividing those functions having connected prisoner sets from those whose prisoner sets are totally disconnected.

Once again, without reference to the value of the parameter, it is possible to determine which of these two outcomes will occur. If iteration of the critical initial point $x = 0$ through $y = x^2 + c$ gives a sequence of values that escapes to infinity, then the prisoner set is disconnected. Otherwise the prisoner set will be an invariant interval.

When the sequence of outcomes generated by iterating the critical point through $y = x^2 + c$ escapes to infinity, we can measure the rate of escape by counting the number of iterations required for the sequence to exceed some fixed large magnitude M.

Iteration Using Complex Numbers

If we allow the value assigned to parameter c is to be a complex number, the main feature to be studied is once again the nature of the prisoner set when x is replaced by domain points that are drawn from the complex plane, called z.

Each setting of the parameter c defines a single function of the form $w = z^2 + c$. For a given function, some points in the complex plane are prisoners while others escape to infinity under iteration. When the parameter value c and the domain are both restricted to real numbers, the collection of all prisoner points for a given function is either a connected invariant interval or a totally disconnected set of points. In the case of complex values for parameter c and a domain of complex points, the prisoner set is also either a connected region or a totally disconnected set of points.

For a given setting of the parameter c, the behavior of the critical point $0 + 0i$ under iteration establishes whether the prisoner set is connected or not. As before, when the critical point does not escape to infinity, the prisoner set is connected. When the critical point escapes to infinity, the prisoner set is totally disconnected. The points that lie on the boundary between the prisoner set and those points that escape to infinity are collectively referred to as the Julia set. When the prisoner set is connected, the Julia set consists of the boundary of this connected set. However, when the prisoner set is disconnected, it has no interior points. Thus, the Julia set is the prisoner set itself.

The Mandelbrot set identifies those parameter values $c = a + bi$ in the function $w = z^2 + c$ for which iteration of the critical point $z = 0 + 0i$ yields a sequence that fails to escape to infinity. Such parameter values are represented visually on the map by black points, and they are associated with connected Julia sets. Points outside of the Mandelbrot set relate to parameter values for which the associated functions iterate the critical point $z = 0 + 0i$ to infinity. These points correspond to disconnected Julia sets, and they are colored according to the rate at which the critical point escapes. The exceedingly irregular boundary of the set of black points forms a barrier between these two types of behavior.

Additional Readings

Chapters 13 and 14 of *Fractals for the Classroom, Part Two,* H.-O. Peitgen, H. Jürgens, D. Saupe, Springer-Verlag, New York, 1992.

USING THE ACTIVITY SHEETS

6.1 Prisoner Versus Escape

Specific Directions. When a pool ball rebounds off the side of a pool table, the angle formed by the path of the ball from the wall equals that which the ball had followed prior to striking the wall. Similarly, the angle of incidence for light rays striking a mirror equals the angle of reflection off the mirror. In this activity the objective is to follow the paths of pool balls and light rays as they traverse back and forth across tables and mirrors of different shapes.

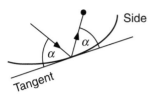

Falling into a pocket is the only way that a ball can escape the bounds of an elliptical pool table. Passing through the focus point of a parabolic mirror is the only way light rays directed into the mirror can escape the mirror. We seek to identify through these examples those conditions that will indicate whether a particular path leads to escape or to forever being a prisoner.

Implicit Discoveries. On an elliptical pool table, a ball can escape the confines of the table if it follows a path leading through a focus. On a parabolic mirror, a light ray can escape the confines of the mirror if its path leads through the focus.

Extensions. Study rectangular pool tables with a single pocket in the center of the table. Is there a critical point on such tables that will determine which paths lead to the pocket?

6.2 Prisoner Sets and Invariant Intervals

Specific Directions. Instead of a local application of graphical iteration to a specific point, consider the collective effect of graphical iteration upon the entire domain of the function. All points from the domain that fail to escape to infinity under iteration form the prisoner set.

Some prisoner sets may contain an entire interval of points from the domain that not only fail to escape to infinity but even fail to escape the bounds of the interval itself. These invariant intervals can be identified from the graph of the function using the box test.

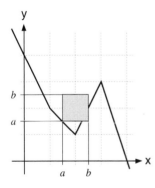

Implicit Discoveries. Marking an interval $[b, c]$ on both the vertical and horizontal axes defines a square within the plane as shown in the figure to the right. If the graph of $f(x)$ neither extends above nor below the square over the interval $[a, b]$, then $[a, b]$ is invariant. Note that this is not satisfied in the figure.

6.3 The Cantor Set

Specific Directions. Begin with a straight line segment and remove the middle third of the segment. Delete the middle thirds from each of the remaining two segments. After infinitely many steps of extracting the middle thirds from all remaining segments, the points that remain are collectively referred to as the Cantor set. For certain functions graphical iteration simulates this construction process.

Implicit Discoveries. For tent functions of the sort shown to the right, whole subintervals escape under graphical iteration from the interval [0,1] leaving only a Cantor set of points as prisoners.

6.4 Invariant Intervals and The Cantor Set

Specific Directions. The graphs of functions of the form $f(x) = ax(1-x)$ with $a > 1$ are parabolas that open downward with x-intercepts at $x = 0$ and $x = 1$. When the vertex of such a parabola lies below the line $y = 1$, the prisoner set is the invariant interval $[0,1]$. However, when the vertex extends above the line $y = 1$, it is possible to identify whole subintervals of $[0,1]$ that escape to (negative) infinity under graphical iteration. Moreover, for each such escaping subinterval S, a large collection of others can be found that iterate into S. Accordingly, this additional collection of subintervals also escapes to infinity through S. The effect is to whittle the prisoner set down to a disconnected dusting of points.

Implicit Discoveries. For each interval that escapes from $[0,1]$ in N iterations, there are two additional intervals within $[0,1]$ that escape from $[0,1]$ in $N+1$ iterations.

6.5 Critical Points

Specific Directions. The objective of this activity is to locate a single critical point within the domain of a quadratic or similar function that singularly discriminates if the prisoner set is an invariant interval or a totally disconnected Cantor set. The approach studies the effect of reducing the maximum height of functions upon those intervals that escape from $[0,1]$ under graphical iteration. Since the maximum height depends upon the value of the parameter a when the function has the form $f(x) = ax(1-x)$, we also seek a barrier value for parameter a that divides those functions having invariant intervals as their prisoner sets from all those that have totally disconnected prisoner sets.

Implicit Discoveries. For parabolas that open downward with x-intercept at $x = 0$ and $x = 1$, the critical point for iteration is $x = 1/2$. The iteration of the critical point reveals whether the prisoner set is the unit interval (the iteration remains in the interval) or totally disconnected (the iteration escapes to infinity).

Extensions. Study other classes of functions such as $f(x) = k\sin x$ on $[0,\pi]$ under graphical iteration for the presence of critical points that determine the nature of the prisoner set.

6.6 Maximal Invariant Interval

Specific Directions. When functions of the form $f(x) = ax(1-x)$ have an invariant interval as the prisoner set, that interval is always $[0,1]$. However, when invariant intervals exist for the class of functions defined by $f(x) = x^2 + c$, the length of the interval increases as the value of parameter c decreases towards -2. The interval of maximum length occurs when parameter c is at the barrier, the value that divides those functions having invariant intervals as their prisoner sets from those that have totally disconnected prisoner sets. This barrier can be found by a trial-and-error approach that geometrically applies the box test to the graph of $f(x) = x^2 + c$ for different values of c. Alternately, one might determine the specific barrier value of parameter c algebraically by analyzing the relative positions of the function and the box that would maximize the invariant interval. Finally, the barrier could be located experimentally by observing the behavior of the critical point $x = 0$ for different values of the parameter c. Irrespective of how it is found, the function $f(x) = x^2 + c$ having the maximal invariant interval also establishes the barrier value for c that separates the two distinct types of prisoner sets for these functions.

Implicit Discoveries. Iterating the critical point $x = 0$ through functions of the form $f(x) = x^2 + c$ singularly determines if the prisoner set is an invariant interval or a totally disconnected Cantor set.

6.7 A Strategic Mapping

Specific Directions. For each function within the $f(x) = ax(1-x)$ class that we have already studied, there is a single related function having similar properties under graphical iteration within the $f(x) = x^2 + c$ class involved in the Mandelbrot set. The identification between functions of the two classes can be defined in terms of a geometric transformation involving a rotation, magnification, and translation, or it can be defined by reference to algebraic formulas. By either approach, for each value assigned to parameter a, there is a corresponding value of parameter c.

Implicit Discoveries. The identification between functions in the class $f(x) = ax(1-x)$ and those in the $f(x) = x^2 + c$ class implies that related functions either both have connected prisoner sets or they have disconnected prisoner sets.

6.8 Rate of Escape

Specific Directions. For some settings of parameter c in $f(x) = x^2 + c$, the critical point $x = 0$, under graphical iteration, fails to escape to infinity. For other settings of parameter c, the same critical point $x = 0$ escapes to infinity. A critical point may be said to escape slowly if a large number N of iterations are required to increase the magnitude of the sequential iterates beyond some large prescribed value. For a different setting of c, the critical point escapes more quickly in the sense that a small number N of iterations rapidly increases the values of the iterates beyond the same prescribed value.

In order to visually map the effect that varying c has upon the critical point, colors are applied when critical points escape according to the number N of iterations. For example, one might assign hot colors to small values of N to indicate large increments and rapid progress outward. Cool colors might be assigned to larger values of N. Although the coloring scheme is arbitrary, the resulting map will not only indicate which parameter values cause a critical point to be a prisoner, but it also indicates something of the rate at which the critical point escapes for other values of c.

Implicit Discoveries. The colored parameter map identifies which values for c lead to connected prisoner sets as well as those values of c that have disconnected prisoner sets.

6.9 Complex Numbers, A New Domain for Iteration

Specific Directions. Complex numbers can be expressed in the form $a + bi$, where a and b are real numbers and $i = \sqrt{-1}$. Addition and multiplication with complex numbers corresponds to performing those same operations using linear binomials in the variable i. In the case of multiplication, simplification of the result follows from the fact that $i^2 = -1$. Accordingly,

$$(a + bi) + (c + di) = (a + c) + (b + d)i$$
$$(a + bi)(c + di) = ac + (ad + bc)i + bdi^2 = (ac - bd) + (ad + bc)i$$

Any complex number $z = x + yi$ can be represented visually on a graph as a point (x, y) or as a vector from the origin to the point (x, y). This representation also allows a geometrical interpretation for addition and multiplication.

Implicit Discoveries. The operations defined for complex numbers allow for their iteration through functions of the form $f(z) = z^2 + c$ since such functions only involve multiplication of a number by itself and addition. Moreover, each complex number $z = x + yi$ corresponds to a single point (x, y) on a graph. Accordingly, the sequence of numbers obtained through iteration may be displayed visually.

6.10 Orbits

Specific Directions. The sequence of iterates obtained by iterating a point z through the function $f(z) = z^2 + c$ are collectively referred to as the orbit of point z. For a given value of parameter c, the orbits of some points z spiral off to infinity when plotted on a graph of the complex plane. For other points z the sequence of iterates remains imprisoned within a bounded region of the plane.

Implicit Discoveries. For some values of parameter c it is difficult to randomly select prisoner points z that have bounded orbits. As shown in subsequent activities, the prisoner set of points associated with such values of c form a totally disconnected dust of points.

6.11 Julia Sets

Specific Directions. For a given setting of parameter c, subject the domain of complex numbers to iteration through the function $f(z) = z^2 + c$. Some complex numbers escape to infinity while those that fail to escape form the prisoner set. Since each complex number $x + yi$ corresponds to a unique point (x, y) on the plane, the collection of points associated with the prisoner set can be graphed. The boundary between the prisoner set and those points that escape to infinity forms the Julia set for the given function. The required complex arithmetic can be performed manually with paper and pencil as shown in the prior activity. However, a calculator or computer program for iterating complex numbers is practically necessary.

Implicit Discoveries. Julia sets are either connected or disconnected. The behavior of the critical point $0 + 0i$ under iteration determines for a given function which of these two possibilities is actually present.

Extensions. Write a program for a computer or graphing calculator that tests points for inclusion in the prisoner set of a given function $f(z) = z^2 + c$. The program should plot points that are within the prisoner set.

6.12 The Mandelbrot Set

Specific Directions. The plot referred to as the Mandelbrot set is, in fact, a plot that shows for each value of c in $f(z) = z^2 + c$ whether or not the associated Julia set is connected. Each value of the parameter $c = a + bi$ is a complex number that corresponds to the point (a, b) on the complex plane. For a graphical representation of the Mandelbrot set we might use a grid to partition the plane into square regions. Selecting a point from each cell, for example the midpoint, or one of the corners, in the grid then also defines a particular function for that region. The square region is colored black only when the critical point $0 + 0i$ is a prisoner under iteration through that defined function. The finer the grid, the more precise the Mandelbrot set is represented.

Implicit Discoveries. Points within the Mandelbrot set correspond to values for parameter c that give rise to connected Julia sets. The Julia sets corresponding to points outside the Mandelbrot set are disconnected.

6.13 Coloring The Mandelbrot Set

Specific Directions. Although the Mandelbrot set is essentially a map of those parameter values c in $f(z) = z^2 + c$ that give rise to connected Julia sets, the use of distinct colors provides a means of mapping additional information on the same plot. Specifically, points outside of the Mandelbrot set correspond to parameter values c in $f(z) = z^2 + c$ for which the critical point iterates to infinity. Such points are colored accordingly to the rate of escape of the critical point.

Implicit Discoveries. The coloring scheme defined above and in the activity sheets yields incredibly intricate patterns on the exterior of the Mandelbrot set.

Extensions. Explore 2-D and 3-D magnifications of the colored Mandelbrot map using the software entitled *The Beauty of Fractals Lab* available from Springer-Verlag. The Mandelbrot set can be programmed on the TI 85 and Casio graphing calculators.

6.1 PRISONER VERSUS ESCAPE   6.1A

A *conic section* is created by slicing through a cone with a plane. Four different two-dimensional shapes are possible.

Every cut through a cone produces a conic section. Cut the cone with a plane perpendicular to its axis and a circle results. Depending on the angle of the cut a hyperbola, a parabola, or an ellipse can be formed if the plane is not perpendicular to the axis.

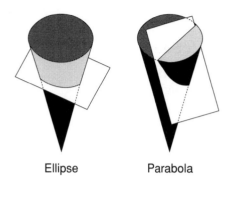

Ellipse   Parabola

The conic sections have been studied for over 2500 years, and they can be viewed geometrically or algebraically. Their equations have distinctive properties as do their corresponding shapes. For example, special points referred to as foci correspond to features that distinguish one conic from another.

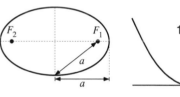
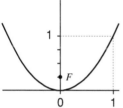

Imagine a mirror that has the shape of a conic section. As rays of light reflect back and forth off its surface, they model an iteration process. In fact, the reflective property of the surface makes the paths of light well-defined and computable.

Reflective property The angle of incidence is equal to the angle of reflection.

On a curved surface, the angles of incidence and reflection are measured from the tangent line at each reflection point.

The mirrors shown to the right have parabolic shapes with sides of unlimited length. Follow the paths of the reflected rays shown.

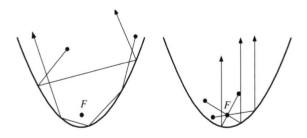

While the focus is not a repelling point in the strict sense, all incoming paths eventually do reflect in an outward direction escaping from the region of the focus. Incoming paths parallel to the axis of the parabola reflect through the focus and escape along other parallel routes as shown.

6.1B

Each of the following parabolas represents a mirror with focus F. Let the point P identify the location of a light source. From each point P, two light rays have been initiated. Where possible, trace their paths through two reflection or iteration stages.

Example   1.   2.

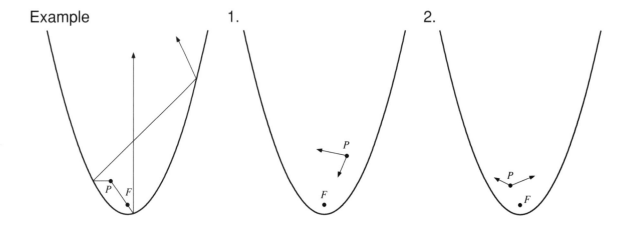

3. If a quality control inspector wants to assess the precision of the mirror's surface, from what single critical point would the direction of the light rays be unimportant, allowing an easy assessment of the reflective behavior of the parabolic mirror?

Imagine a game of pool played on a table having an elliptical shape. The ball rebounding back and forth off the curve gives another model of the iteration process. The same reflective property applies as with light bouncing off a mirror.

On an elliptical pool table, there are no corners for the pockets as in a regular rectangular pool table. Instead, suppose a single pocket is placed at one of the two foci and a mark placed at the other. An accurate shot would send the ball directly into the pocket. Surprisingly, shooting the ball in any direction at all from the marked focus point also causes the ball to rebound from the perimeter and go directly into the pocket. This is a unique property of an ellipse, and it implies that shooting the ball from any point through the marked focus results in its falling into the pocket, thereby escaping the confines of the table.

Escape paths on an elliptical pool table are those that contain the pocket or the marked focus. Those paths that do not contain the foci imprison the ball forever within the walls of the table. Such paths are called prisoner paths.

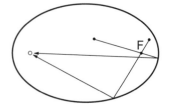

Escape paths

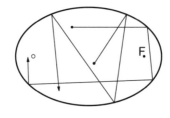

Prisoner paths

On each elliptical pool table shown, one focus is marked F with a round pocket drawn at the other. From each point P, draw a line through the marked focus F to the wall of the table. Then draw a line representing the path of the ball after it rebounds from the wall and travels towards the pocket.

Example   4.   5.

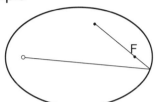 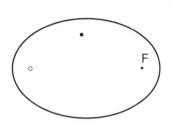 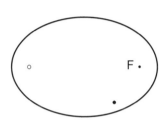

6. Assume a random starting point is selected on an elliptical pool table. The shot must be aimed carefully towards one of two special locations in order to insure that the ball lands in the pocket. Where are those two points?

7. Suppose a quality control inspector wants to assess the table's behavior. From what single *critical point* would the direction of the shot be unimportant, thereby allowing an easy check of the behavior of the elliptical pool table?

Assume that a special ball placed on an elliptical pool table has the unusual property that, once struck, its stages of iteration or reflection never stop unless it escapes the table by falling into the pocket. Show why the paths initiated on the tables below never escape by leading to the pockets.

8.   9.

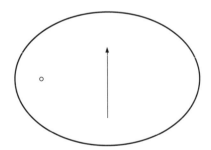 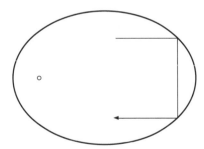

10. Suppose the path of a ball on an elliptical pool table eventually retraces itself. In such a case, the ball returns to points along the perimeter in some repetitive, iterative pattern. If we refer to the number of points in the pattern as the period of the path, what periods are represented by the paths shown in questions 8 and 9?

6.1D

11. Can an escape path on an elliptical pool table have more than one iteration or reflection off the surface?

All iteration paths of a ball in the elliptical pool table exhibit one of two basic properties. They are either *prisoners* entrapped by the ellipse or they escape, freed from the curves that otherwise surround them. Under certain conditions, the trapped paths periodically retrace themselves. However, whether a path escapes or is imprisoned, a single *critical point* exists that may be used to determine the behavior of the paths within the given curve.

A geometric approach was used to develop the initial ideas given in this activity. Comparable questions can be asked in an algebraic form.

The equation $\frac{x^2}{25} + \frac{y^2}{16} = 1$ defines an elliptical pool table with foci at (3,0) and (-3,0).

12. Which of these linear equations locate paths of the pool ball that lead to the pocket at (-3,0)?

   a. $y = 0$ b. $y = -x + 3$ c. $y = x - 3$

13. Write the equation of a line through each point locating a path that rebounds off the wall of the pool table and escapes through the pocket at (-3,0).

   a. (0,0)   b. (2,2)   c. (0,3)

14. A pool ball is shot from (0,-4), rebounds off the wall once, and then goes into the pocket at (-3,0). At what point does it hit the elliptical wall of the pool table?

The equation $y = \frac{x^2}{4}$ defines a parabolic mirror with focus at (0,1).

15. Which of these linear equations locate reflecting rays through the focus that escape along lines parallel to the axis of the parabola?

   a. $x = 0$ b. $y = \frac{x}{2} + 1$ c. $y = -2x$

16. Write the equations of two lines through each point which locate rays that reflect off the mirror in paths parallel to the axis of the parabola.

   a. (2,4)   b. (-3,6)   c. (-0.5,1.5)

6.2 PRISONER SETS AND INVARIANT INTERVALS   6.2A

Some points in the domain of a function under graphical iteration produce sequences of iterates that escape to infinity. Other domain points yield sequences that fail to escape to infinity. These latter points are collectively referred to as the prisoner set P.

Consider the piecewise defined function below with its accompanying graph.

$$f(x) = \begin{cases} 2x & \text{when} & x \le 1 \\ x + 1 & \text{when} & 1 < x \le 2 \\ -2x + 7 & \text{when} & 2 < x \le 3 \\ 3x - 8 & \text{when} & 3 < x \end{cases}$$

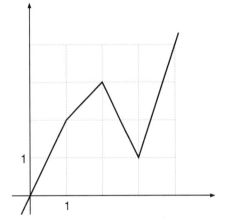

1. Substitute into the piecewise defined function $f(x)$ above in order to compute the values that would be obtained if graphical iteration were applied to the initial points given in the column labeled x_0.

x_0	$x_1 = f(x_0)$	$x_2 = f(x_1)$	$x_3 = f(x_2)$	$x_4 = f(x_3)$	Prisoner
−0.5	−1	−2	−4		
2.75					
3.0					
4.5	5.5	8.5			

2. Write NO in the prisoner column if the point escapes to positive or negative infinity and thereby is not a part of the prisoner set P.

If the interval defined by $f([b, c])$ is a subset of interval $[b, c]$, then interval $[b, c]$ is called *invariant*. Written symbolically, $f([b, c]) \subset [b, c]$ implies that each point in $[b, c]$ is mapped under graphical iteration to a point that is also within $[b, c]$. Such intervals are contained in the prisoner set P of function f since points in such intervals fail to escape to infinity. In other words, interval $[b, c]$ is invariant, and therefore a part of the prisoner set P of the function f, if $f(x) \in [b, c]$ whenever $x \in [b, c]$.

Each of the following intervals $[b, c]$ fails to be invariant under graphical iteration when using the function f as defined above. For each interval, identify a point x in $[b, c]$ for which $f(x)$ is not in $[b, c]$. That is to say, find an x such that $x \in [b, c]$ but $f(x) \notin [b, c]$.

3. $[0, 1]$ 4. $[3, 4]$ 5. $[2, 3]$

6.2B

The range of a function over a given domain consists of all values assumed by the function over that domain.

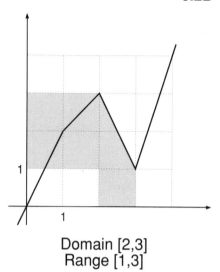

Determine the range of function f for each of the following domain intervals. If the domain and associated range intervals were plotted onto the same number line, in which cases would the range interval be a subset of the domain interval? Explain why intervals having this property are invariant under graphical iteration.

6. $[0, 2]$ 7. $[0, 3]$ 8. $[1, 4]$

Domain [2,3]
Range [1,3]

Associated with the domain interval $[0, 4]$, the range of the function f is also the interval $[0, 4]$. This implies that each of the intervals given in questions 6–8 are contained in the prisoner set P of function f even though some of these intervals fail to be invariant.

On any graph, the notation $[p, q] \times [r, s]$ defines a rectangle bounded by the vertical lines $x = p$ and $x = q$ and by the horizontal lines $y = r$ and $y = s$.

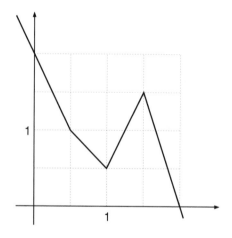

9. Sketch the two squares defined by $[b, c] \times [b, c]$ onto the graph at the right.

    a. $b = 0$ and $c = 1$
    b. $b = 0.5$ and $c = 1.5$

10. In which case in question 9 does the square $[b, c] \times [b, c]$ contain all of $[b, c] \times f([b, c])$?

Points which are prisoners never escape to infinity. Sometimes these points are part of invariant intervals. Such intervals can be located by the following box test.

> *Box Test* If the rectangle $[b, c] \times f([b, c])$ is inside the square $[b, c] \times [b, c]$, then the interval $[b, c]$ is invariant.

11. Use the box test to locate the largest invariant interval shown in each of the two graphs displayed in this activity.

6.3 THE CANTOR SET

6.3A

Much of the current work in fractals and chaos is built on ideas first generated by Georg Cantor (1845–1918) in his work on set theory. One of these fundamental, underlying concepts is the Cantor set.

Begin with a straight line segment having a length of one unit. Remove the middle third, as shown in stage 1, forming two segments of length one-third. Remove the middle third of each of these two segments, and four segments of length one-ninth are formed. Continue ad infinitum this iterative process of removing the middle third of every segment formed at each and every stage. The number of segments increases without bound while their lengths decrease to zero. No intervals remain in the final state, only a dust of points. The final resulting infinite set of disconnected points is the classical Cantor set.

1. The first stage in constructing the Cantor set is shown. On successive lines, construct the second, third, and fourth stages.

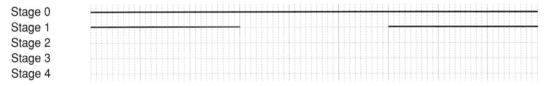

2. How many subintervals are in stage 5? in stage 6? in stage n?

3. What is the length of each segment in stage 5? in stage 6? in stage n?

Graphical iteration can also lead to the construction of the Cantor set. Consider a tent function where the middle third of the function extends above the line $y = 1$. As shown in stage 1 below, initial points selected between 1/3 and 2/3 will escape from the interval $[0, 1]$ in one step of graphical iteration.

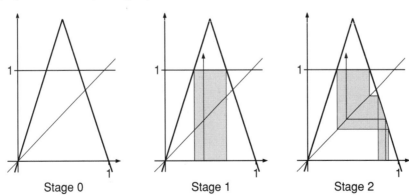

4. The stage 2 graph shows that initial points in the interval between 7/9 and 8/9 escape from the interval $[0, 1]$ in two steps of graphical iteration. Find another interval between 0 and 1/3 that also escapes in two steps of graphical iteration. Shade in the escape path that this interval follows.

6.3B

5. The iteration of the boundary points 7/9 and 8/9 shows how the interval iterates into the interval between 1/3 and 2/3 and escapes.

$$7/9 \rightarrow 2/3 \qquad 8/9 \rightarrow 1/3$$

Give the numerical values of the interval boundary points found in question 4 and show how they too lead to an interval that escapes.

6. Use your results from question 1 to mark on the dotted line below the graph to the right those intervals that remain at stage 3 of the Cantor set.

7. Use graphical iteration to shade in the escape paths for two new escape intervals, one between 0 and 1/9, and the other between 2/3 and 7/9.

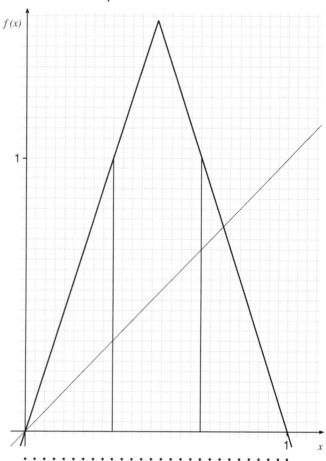

8. The interval between 7/27 and 8/27 escapes by this iteration of its end points:

$$7/27 \rightarrow 7/9 \rightarrow 2/3 \qquad 8/27 \rightarrow 8/9 \rightarrow 1/3$$

Show how the interval between 25/27 and 26/27 escapes in a similar fashion.

9. By two steps of graphical iterations, three subintervals of $[0, 1]$ have escaped. How many subintervals escape after three steps of graphical iteration? after four steps?

10. Write a formula relating the number E of escaping subintervals to the graphical iteration stage N. What happens to E as N increases towards infinity?

6.4 INVARIANT INTERVALS AND THE CANTOR SET   6.4A

Intervals are connected sets of points, while the Cantor set is a disconnected set of points. Depending upon the value of parameter a, graphical iteration performed on a parabola of the form $f(x) = ax(1 - x)$ can produce both prisoner sets that are invariant intervals as well as those that are disconnected Cantor sets.

From an intuitive point of view, a set is *disconnected* if it can be partitioned into parts that do not touch each other. The Cantor sets formed by graphical iteration and mentioned above are disconnected since whole intervals are excluded in their construction. In fact, between any two points of a Cantor set there lies an interval which does not belong to the Cantor set. Therefore, the Cantor set is called *totally disconnected*. We use this intuitive notion of broken versus unbroken to differentiate between the disconnected dust of the Cantor sets and the connected prisoner sets.

Show which of the three graphs have values of x in $[0, 1]$ for which $f(x)$ is *not* in $[0, 1]$. In other words, identify points $x \in [0, 1]$ for which $f(x) \notin [0, 1]$.

1.   2.   3.

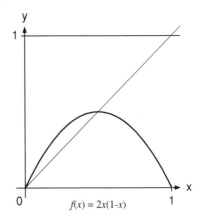

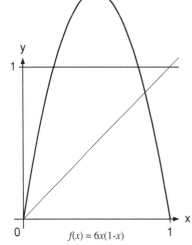

$f(x) = 2x(1-x)$ $f(x) = 4x(1-x)$ $f(x) = 6x(1-x)$

4. The functions shown in the table below correspond to those graphed above. Use their equations to compute successive values through numerical iteration, starting with the initial values supplied in the column labeled x_0.

Function $f(x)$	x_0	$x_1 = f(x_0)$	$x_2 = f(x_1)$	$x_3 = f(x_2)$
$f(x) = 2x(1 - x)$	−0.5	−1.5	−7.5	−127.5
	1.5			
$f(x) = 4x(1 - x)$	−0.5			
	1.5	−3.0	−48.0	
$f(x) = 6x(1 - x)$	−0.5			
	1.5	−4.5	−148.5	
	0.9	0.54		

5. For each of the functions in the table, explain why none of the given points is a member of the associated prisoner set.

Invariant intervals exist for the two functions graphed above in questions 1 and 2. In fact, for both of these two functions, the prisoner set consists of nothing more than the maximal invariant interval.

6. Use the box test on each of these two graphs to find the maximal invariant interval $[b, c]$ such that $f([b, c])$ is a subset of $[b, c]$.

7. Are the prisoner sets defined by the invariant intervals connected or disconnected?

When the vertex of the parabola given by $f(x) = ax(1 - x)$ reaches above the horizontal line $y = 1$, it extends beyond the square $[0, 1] \times [0, 1]$. In such cases the box test indicates that no invariant intervals exist. The function $f(x) = 6x(1 - x)$ shown in question 3 above provides an example of such a parabola. The remainder of this activity explores the nature of prisoner sets for those quadratic functions $f(x) = ax(1 - x)$ where the parameter a is chosen so that no invariant interval exists.

For the particular function of the form $f(x) = ax(1 - x)$ graphed below, each point in the interval $(b, c) = (0.45, 0.55)$ is mapped outside of $[0, 1]$ in one step of graphical iteration. As shown in the graph to the left, the smaller interval (d, e) is mapped by one step of graphical iteration into (b, c) and therefore outside of $[0, 1]$ in the second application of graphical iteration.

8. On the graph to the right, shade the escape path for points in the interval (f, g). The function is $f(x) = \frac{400}{99}x(1 - x)$.

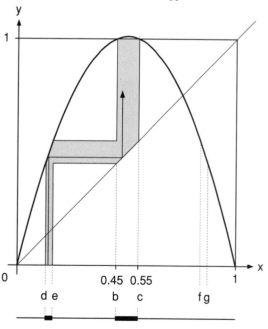

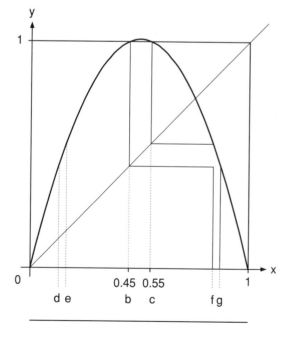

6.4C

9. On the line below the graph to the left on the opposite page, the bold portions mark the two identified intervals (d, e) and (b, c) that map outside of $[0, 1]$ under graphical iteration. Use the line below the graph to the right to mark in a similar fashion the intervals (b, c), (d, e), and (f, g) that map outside $[0, 1]$.

10. Use the fact that $f(d) = b$ and $f(e) = c$ to compute and numerically label the end points of the interval (d, e). Then repeat the process for the end points of (f, g).

In the graphs on the opposite page, horizontal line segments are extended from the points where the vertical lines $x = b$ and $x = c$ intersect the diagonal line $y = x$. The intersection points of these horizontal line segments and the parabola to the left locate the end points of the interval (d, e). Extending the horizontal segments to the right locate the end points of the interval (f, g).

11. On the graph displayed on the side, locate and add vertical lines so as to define two new intervals that map into $[d, e]$ under graphical iteration. Find another two that map into $[f, g]$ under graphical iteration. Since these new intervals map into existing escape intervals, they too must iterate points out of the interval $[0, 1]$. Again the function is $f(x) = \frac{400}{99} x(1 - x)$.

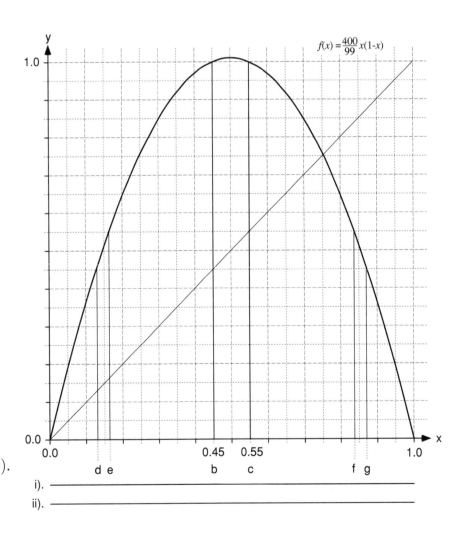

$f(x) = \frac{400}{99} x(1-x)$

i). _____

ii). _____

12. Extend all vertical lines in the graph of question 11, including both $x = 0$ and $x = 1$, through the two lines (i) and (ii) below it. Use bold line segments to mark the four new subintervals as well as the intervals (b, c), (d, e), and (f, g) on the first line (i).

The points within the marked intervals on line (i) are not a part of a prisoner set since they are mapped outside of $[0, 1]$.

13. On line (ii), use bold segments to mark those intervals not marked on line (i).

All prisoner points must lie within the marked intervals on line (ii). However, not all points within these marked intervals are in the prisoner set. Additional stages of iteration will remove additional intervals of points.

14. Lines (i) and (ii) exhibit intervals as they appear after applying three steps of graphical iteration to all of the points in $[0, 1]$.

    a. How many intervals would appear on the first line segment (i) above having the property that they escape from $[0, 1]$ in fewer than five iterations?

    b. How many intervals would appear on the second line segment (ii) above having the property that they fail to escape from $[0, 1]$ in fewer than five iterations?

15. Show that the end points of the intervals identified with the points b, c, d, e, f, and g are all prisoners in $[0, 1]$ by determining the single value to which they would all ultimately iterate after many applications of graphical iteration.

Clearly the maximum height of the function $f(x) = ax(1 - x)$ determines the nature of the prisoner set that results from an endless number of graphical iterations. As the height increases above the line $y = 1$, the prisoner set makes an abrupt, striking change from the interval $[0, 1]$ to a totally disconnected Cantor set. Each different function having maximum height greater than $y = 1$ generates a distinct Cantor set.

The results are summarized below.

For functions of the form $f(x) = ax(1 - x)$ with parameter $a \geq 1$, the prisoner set consists of:

• a connected set when the maximum height lies between $y = 1/4$ and $y = 1$. This connected set is the invariant interval $[0, 1]$.

• a totally disconnected set of points when the maximum height exceeds $y = 1$. This is a Cantor set, a dust of points that contains no intervals.

6.5 CRITICAL POINTS   6.5A

We have been studying functions that have the form $f(x) = ax(1 - x)$ where $a \geq 1$. For such functions, the prisoner set contains no points outside of the interval $[0, 1]$. Depending upon the value assigned to parameter a, the prisoner set either consists of the entire invariant interval $[0, 1]$ or a set of disconnected points from within $[0, 1]$. This activity identifies a simple means for predicting which of these two behaviors occurs.

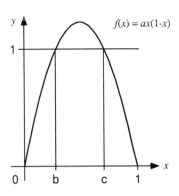

The figure above displays an interval (b, c) centered at $x = 0.5$. Points within this interval (b, c) all leave the interval $[0, 1]$ after only one application of graphical iteration. Subsequent iterations of these points result in their escaping to infinity.

1. For each of the tent functions graphed to the right, draw vertical lines to define an interval of points (b, c) that would leave the interval $[0, 1]$ after only one step of graphical iteration.

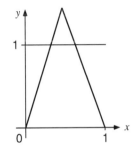 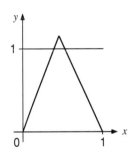

Observe how the maximum value of the tent function $f(x)$ affects the width of your intervals (b, c) in the curves shown for question 1.

2. As the maximum value of $f(x)$ decreases, what happens to the width of the escape interval (b, c)?

Absolute value curves such as those shown in question 1 have equations of the form $f(x) = a(0.5 - |x - 0.5|)$.

3. What is the maximum height of the function $f(x) = 6(0.5 - |x - 0.5|)$? What are the boundaries of the escape interval (b, c) within $[0, 1]$? What is its width?

4. What maximum height is needed for the escape interval (b, c) of absolute value functions of this type to shrink down to the single point $x = 0.5$?

5. Find the maximum value of the parameter a in $f(x) = a(0.5 - |x - 0.5|)$ needed for the absolute value curve to have the property that $x = 0.5$ fails to escape to infinity? Use the information found in question 4.

6.5B

6. The parabolas to the right have equations of the form $f(x) = ax(1 - x)$. For each parabola $f(x)$, draw vertical lines to define an interval of points (b, c) that would leave the interval $[0, 1]$ after only one step of graphical iteration.

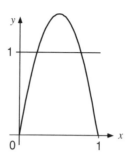 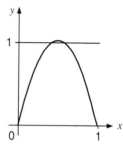

7. What maximum height or y-coordinate of the vertex of these parabolas is needed for the interval (b, c) to shrink down to the single point $x = 0.5$?

8. What maximum value of the parameter a in $y = ax(1 - x)$ is needed for the parabola to have the property that $x = 0.5$ fails to escape to infinity? Compute a by letting y equal the maximum height determined in question 7 using $x = 0.5$.

A parameter's value is referred to as a *barrier* if it is on the boundary between two distinctly different types of behavior. The boundary values found for the parameter a in questions 5 and 8 are such barriers.

9. Write a rule that describes the behavior of graphical iteration when values are selected for the parameter a that are less than the barrier value. What effect upon graphical iteration comes from parameter values greater than that of the barrier?

For functions of the form $f(x) = ax(1 - x)$ or $f(x) = a(0.5 - |x - 0.5|)$, there is only one *critical initial point* that can be used to determine if any points exist within $[0, 1]$ that escapes from that interval under graphical iteration.

10. For $f(x) = a(0.5 - |x - 0.5|)$, explain why the critical initial x-coordinate is $x = 0.5$. What is the critical initial x-coordinate for $f(x) = ax(1 - x)$?

The critical initial point x_0 for graphical iteration through $f(x)$ occurs on the x-axis at the x-coordinate of the function's maximum point. The nature of the prisoner set can be determined from that single point. If the curve is above the line $y = 1$ at that point, the initial point escapes and thereby indicates that the prisoner set is a Cantor set.

As an optional extension requiring calculus, consider the graphs of functions defined by $f(x) = ax(1 - x^2)$ for $x \geq 0$ and $f(x) = ax$ for $x < 0$. Such functions look somewhat similar to parabolic graphs. However, unlike the graphs of $f(x) = ax(1 - x)$, they fail to exhibit symmetry about vertical lines through their maximum points. This lack of symmetry complicates finding the critical point. Fortunately, calculus provides methods for finding relative maximum points on the graph of a function such as this.

11. Consider the function $f(x) = ax(1 - x^2)$ for $x \geq 0$ and $f(x) = ax$ for $x < 0$. Use calculus to locate the relative maximum point of $f(x)$ within $[0, 1]$. What critical initial point within interval $[0, 1]$ can be used to determine if any points exist within $[0, 1]$ that escape to infinity under graphical iteration of the function ?

6.6 MAXIMAL INVARIANT INTERVAL   6.6A

In Activity 6.2, the following box test was formulated for locating intervals that remain invariant under iteration through a function f.

BOX TEST   If the rectangle $[b, c] \times f([b, c])$ is inside the square $[b, c] \times [b, c]$, then the interval $[b, c]$ is invariant.

In this activity, we will identify the function of the form $f(x) = x^2 + c$ that has the largest invariant interval.

1. Use the box test centered at the origin in order to locate on each graph the largest possible invariant interval. In each case draw the box and highlight the invariant interval.

a. $f(x) = x^2 - 0.25$

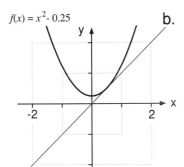

b. $f(x) = x^2 - 0.75$

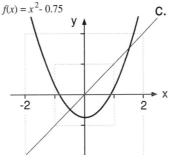

c. $f(x) = x^2 - 2$

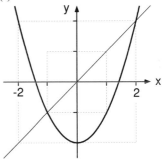

The preceding results suggest that decreasing the value of the parameter c translates the graph of $f(x) = x^2 + c$ downward and increases the size of the invariant interval.

The boxes sketched for question 1 should all have one corner located at the intersection point of $f(x) = x^2 + c$ and the diagonal $y = x$. The sides of the boxes intersect the x-axis and act as bounds for the invariant intervals.

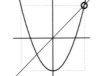

2. The box displayed on the graph to the right defines an interval on the x-axis that fails to be invariant. Explain why the resulting prisoner set is a totally disconnected Cantor set.

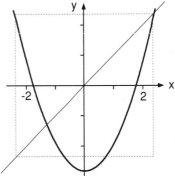

The invariant interval of maximum length occurs when the vertex of the parabola is on the base of the box centered at the origin with one corner at the intersection point of $f(x) = x^2 + c$ and the diagonal $y = x$. Questions 3 and 4 supply an algebraic procedure for computing that value for c that will place the parabola in this strategic position.

3. Suppose the y-intercept of $f(x) = x^2 + c$ is below the x-axis. If the function $f(x)$ and the diagonal $y = x$ intersect, show that the x-coordinate of the intersection point in the first quadrant is given by $x = 1 + \sqrt{1 - 4c}$. Hint: Begin with $x^2 + c = x$.

4. Suppose the y-intercept of $f(x) = x^2 + c$ is below the x-axis. What value must be assigned to c so that the vertical distance from the origin down to the y-intercept of $f(x)$ equals the horizontal distance from the origin to $x = 1 + \sqrt{1 - 4c}$. Hint: Use $f(x) = x^2 + c = x$ with $x = -c$.

Use the programs given below to experimentally determine the value of the parameter c in $f(x) = x^2 + c$ that acts as a barrier between those functions having an invariant interval as the prisoner set and those for which the prisoner set is a disconnected Cantor set.

Line	CASIO	Line	TEXAS INSTRUMENTS
1		1	:ClrHome
2	0 → R	2	:0 → R
3	0 → X	3	:0 → X
4		4	:Disp "C="
5	"C="? → C	5	:Input C
6	" "	6	:Disp " "
7	Lbl 1	7	:Lbl 1
8	X △	8	:Disp X
9		9	:Pause
10	X²+C → X	10	:X²+C → X
11	R+1 → R	11	:R+1 → R
12	R<8 =>Goto 1	12	:If R<8
13		13	:Goto 1

Lines: 1–3   initialize the counter R and fix the initial point at $x = 0$.
        4–6   allow for input of parameter c.
        7–13   provide the iteration loop for eight steps of graphical iteration.

5. Enter the appropriate program into your graphing calculator. The programs perform iteration for the parabolic function $f(x) = x^2 + c$ from the initial value of $x = 0$, the critical point for parabolas of this sort. Repeatedly run the program entering values for c between $c = -1$ and $c = -4$. Find the value for c that is the barrier between those function for which the critical point iterates to infinity and those functions for which the critical point does not iterate to infinity.

The parameter value $c = -2$ in $f(x) = x^2 + c$ acts as a barrier separating those functions having an invariant interval as the prisoner set from those for which the prisoner set is a disconnected Cantor set of points.

6.7 A STRATEGIC MAPPING   6.7A

Like the function $f(x) = ax(1-x)$ studied in previous activities, those having the form $g(x) = x^2 + c$ are also quadratic. In fact, for each function of the form $f(x) = ax(1-x)$, there is a related function $g(x) = x^2 + c$ having similar properties that can be found geometrically by a particular rotation, translation, and magnification. These latter functions form the class used for generating the Mandelbrot set.

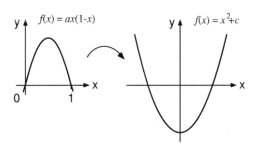

Study the sequence of three transformations applied to triangle P below involving a magnification factor $a = 2$.

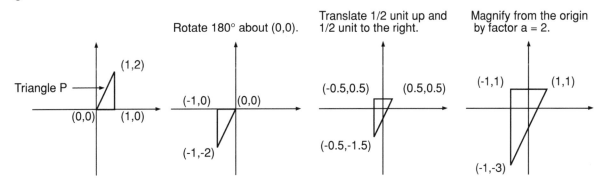

Rotate 180° about (0,0).   Translate 1/2 unit up and 1/2 unit to the right.   Magnify from the origin by factor a = 2.

1. Draw the results of applying the same transformations as those shown above to the vertices of triangle Q. This time use a magnification factor of $a = 4$. Label the coordinates of the vertices after each step in the sequence.

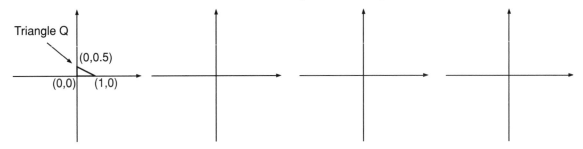

The sequence of geometric transformations just practiced may be accomplished algebraically by applying the following formulas for computing image coordinates (x', y') from the given original coordinates (x, y):

$$x' = -ax + \frac{a}{2} \quad y' = -ay + \frac{a}{2}$$

Observe that the formulas for transforming x into x' and y into y' are the same.

6.7B

2. Use the algebraic equations with the given values for a. Then substitute the coordinates (x, y) for the vertices of the given triangle into the equations in order to determine the vertices (x', y') of the image triangle.

$a = 3$ $x' = -3x + 3/2$ $a = 4$ $x' = -4x + 2$
           $y' = -3y + 3/2$ $y' = -4y + 2$

Triangle P Defined by Vertices	Vertices of Image Triangle		Triangle Q Defined by Vertices	Vertices of Image Triangle
(0,0)	(1.5,1.5)		(0,0)	(2,2)
(1,0)			(2,0)	
(1,2)			(0,1)	

3. The parabola $y = 3x(1 - x)$ has its vertex at $(0.5, 0.75)$ and passes through the points $(0, 0)$ and $(1, 0)$.

Use the transformation equations given above with $a = 3$ to find the image of each of these three points.

Then sketch a graph of the new parabola that passes through all three of these image points. It is the parabola $y = x^2 + 0.75$. Its width is $a = 3$ (x-intercepts at $-a/2$ and $a/2$) and its height (y-intercept) is $(a/2)^2 = 9/4$.

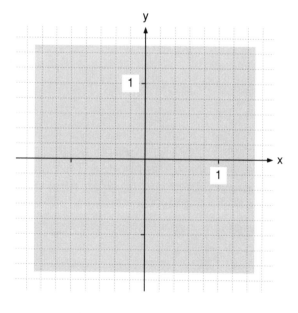

4. Let us investigate the transformation of the parabola in detail. Use the graphs provided on the following page.

a. This is the graph of the function $y = 3x(1 - x)$ for $x \in [0, 1]$.

b. Draw a graph showing the result of rotating the parabola 180 degrees about $(0, 0)$.

c. Translate the parabola obtained in part (b) 0.5 units up and 0.5 units to the right.

d. Draw a graph of the function obtained by magnifying the parabola of part (c) by a factor of 3 units from (0,0).

6.7C

a.

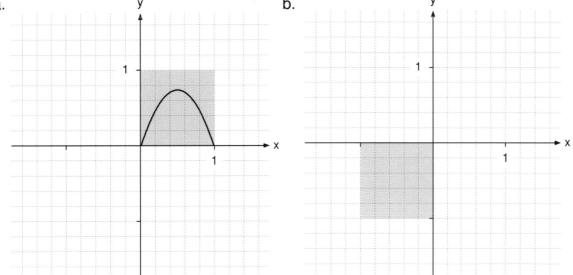

b.

c.

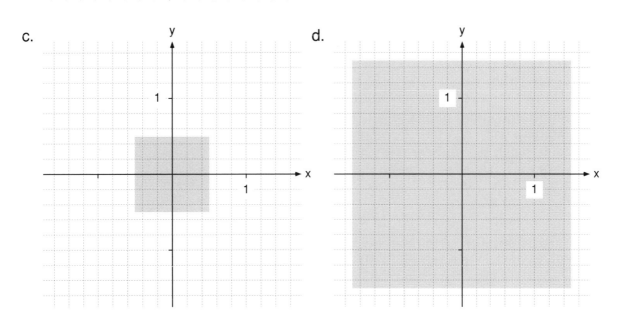

d.

The rotation by 180 degrees changes the original parabola into one that opens to the top. The translation positions the vertex onto the y-axis. Note that the height of the shown parabola is $a/2$ and its width 1. Finally, the magnification turns the curve into a standard parabola of the form $y = x^2 + c$.

5. What is the height and width of the final parabola in question 4d? What is the y-coordinate of the vertex. Compare your graphs from questions 3 and 4d.

6.7D

For each function of the form $f(x) = ax(1 - x)$, a corresponding function of the form $g(x) = x^2 + c$ can be found by applying a series of transformations to the graph of f:
- the composition of a 180 degrees rotation about the origin,
- a translation of 1/2 units up and 1/2 units to the right, and
- a magnification by the factor a from the origin.

For the latter function, the constant c can be computed as $c = \frac{a}{2} - \frac{a^2}{4}$.

6. Each of the following functions represents an original parabola of the form $y = ax(1 - x)$. Determine the equation of the image parabola $y = x^2 + c$ obtained by mapping each original through the transformation equations, $x' = -ax + a/2$ and $y' = -ay + a/2$.

Original	$y = 1x(1 - x)$	$y = 3x(1 - x)$	$y = 4x(1 - x)$	$y = ax(1 - x)$
Image				$y = x^2 + \left(\frac{a}{2} - \frac{a^2}{4}\right)$

7. Perform two steps of graphical iteration at the original parabola in 4a. Use $x_0 = 0.3$ as initial value. Then apply the three transformations to the grapical iteration path. Draw the rotated path into 4b, the rotated and translated one into 4c, and the completely transformed path into 4d.

8. Do two steps of graphical iteration on the transformed parabola $y = x^2 - 0.75$ starting with $x'_0 = 0.6$. Compare the result with the transformed parabola and path drawn in exercise 4d. Observe that $x' = -ax + a/2$ transforms $x_0 = 0.2$ into $x'_0 = 0.9$.

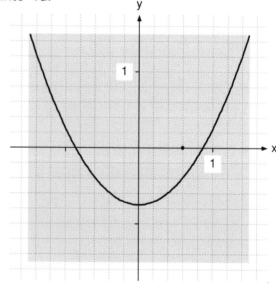

The transformation of the parabola $f(x) = ax(1 - x)$ into the parabola $g(x) = x^2 + c$ also maps the iteration of an initial value x_0 under $f(x)$ into the iteration of $x'_0 = -ax_0 + a/2$ under $g(x)$. In this sense, the two parabolas are equivalent.

The above equations also map the prisoner set of $f(x)$ to the prisoner set of $g(x)$.

9. The interval $[0, 1]$ is the prisoner set of $y = ax(1 - x)$ when $1 \le a \le 4$. Find the images of the end points of $[0, 1]$, recording the interval defined by the x-coordinates of these images in the table below.

Original	$y = 1x(1 - x)$	$y = 3x(1 - x)$	$y = 4x(1 - x)$	$y = ax(1 - x)$
Prisoner set	$[0, 1]$	$[0, 1]$	$[0, 1]$	$[0, 1]$
Image	$\left[-\frac{1}{2}, \frac{1}{2}\right]$			

6.8 RATE OF ESCAPE   6.8A

When the prisoner set is an invariant interval, the sequence generated by iterating the critical point $x = 0$ never escapes the interval. However, when the prisoner set is a Cantor set of disconnected points, the sequence generated by the critical point $x = 0$ does escape to infinity. By selecting some large fixed magnitude as a milepost, the number of terms required for the sequence of iterates to step past this marker measures the rate at which such initial points escape.

The programs listed below perform iteration for the parabolic function $f(x) = x^2 + c$. Starting from the critical initial point $x = 0$, each program counts the number of iterations required before values in the sequence of iterates exceed 1000.

Line	CASIO	Line	TEXAS INSTRUMENTS
1	$0 \rightarrow N$	1	$:0 \rightarrow N$
2	$0 \rightarrow X$	2	$:0 \rightarrow X$
3	Lbl 1	3	:Lbl 1
4		4	:ClrHome
5		5	:Disp "ENTER C<-2"
6	"C="? $\rightarrow$ C	6	:Input C
7	C>=-2 => Goto 1	7	:If C ≥ -2
8		8	:Goto 1
9	" "	9	:Disp " "
10	Lbl 2	10	:Lbl 2
11	N+1 $\rightarrow$ N	11	:N+1 $\rightarrow$ N
12	$X^2 + C \rightarrow X$	12	:$X^2 + C \rightarrow X$
13	X$<$1000 => Goto 2	13	:If X$<$1000
14		14	:Goto 2
15	N$\triangle$	15	:Disp N
16		16	:End

Lines: 1–2   Initialize the counter N and fix the initial point at $x = 0$.
         4–9   Allow for input of the constant parameter c.
       10–16 Form the loop which counts the number of iterations required before the values of successive iterates exceed 1000.

1. Enter the appropriate program into your graphing calculator.

2. Complete the table by using one of the above programs. For each entered value of parameter c, the program will determine the number N of iterations before the subsequent values exceed 1000.

Value of c	-3	-2.1	-2.01	-2.001	-2.0001	-2.00001	-2.000001
Iterations N							

6.8B

3. On the axes to the right, sketch a graph of the data that you recorded in the table for question 2. Based on the assumption that the data can approximated by a line of best fit what is the quantity to which the number of iterations is proportional?

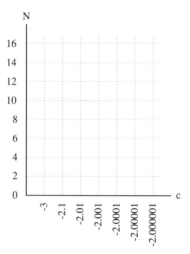

4. In Activity 6.6 we determined that $c = -2$ acts as a barrier between those functions that have $-2 \leq c < 1/4$ and possess invariant prisoner intervals and those that have $c < -2$ and do not possess prisoner intervals.

   For those functions that do not possess prisoner intervals, use your graph to explain the effect of selecting values for parameter c that increase towards the barrier of -2 as the iteration count increases.

Rate of escape of critical point   Let $f(x) = x^2 + c$ be a function involving the parameter c. Select M to be a large and fixed magnitude and compute the iteration of $f(x)$ started at the critical point $x_0 = 0$. The *escape rate of the critical point* is defined to be the minimal number of iterations necessary for the iterates to exceed M in absolute value.

5. The horizontal axis of the graph above is represented in the chart below. As before, the listed numbers represent some of the possible parameter values for c in $f(x) = x^2 + c$. For each value of c identified with a subinterval on the axis below, use the value of N which you obtained in question 2 in order to assign a color to the subinterval.

0–5	6–7	8–9	10–11	12–13	14–15
yellow	orange	brown	red	green	blue

Fill in each subinterval cell with the appropriate color to obtain a visual map of the escape times associated with the various values of c.

−3	−2.1	−2.01	−2.001	−2.0001	−2.00001	−2.000001

When the parameter c in $f(x) = x^2 + c$ is assigned values that give rise to disconnected Cantor sets, the value of c can be given a color so as to visually represent the rate at which the critical point escapes to infinity.

6.9 COMPLEX NUMBERS, A NEW DOMAIN FOR ITERATION   6.9A

When constructing the Mandelbrot set, iteration is performed upon complex numbers. The complex number system consists of all elements that can be expressed in the form $x + yi$ using $i = \sqrt{-1}$ with x and y both real numbers. Many students first encounter complex numbers when solving equations using the quadratic formula. In this activity, some of the important properties of complex numbers are developed. Note initially that $i = \sqrt{-1}$ implies $i^2 = -1$.

In a complex number of the form $x + yi$, the real number x is referred to as the real part and the real number y as the imaginary part. We accomplish addition of complex numbers $a + bi$ and $c + di$ by adding the corresponding parts to obtain the complex number $(a + c) + (b + d)i$. For example,

$$(1 + 4i) + (3 + 2i) = (1 + 3) + (4 + 2)i = 4 + 6i$$

1. Add the following complex numbers.

    a. $(2 + i) + (1 + 2i)$ _____   b. $(2 - 2i) + 3i$ _____

    c. $(3 - 4i) + (1 + i)$ _____   d. $(-3 - 2i) + (1 + 4i)$ _____

Since every complex number involves two real numbers x and y, we can visually represent each complex number $x + yi$ on the coordinate plane by plotting the point (x, y). In fact, if you review your results to questions 1 and 2, you can observe that both the product and the sum of two complex numbers is also a complex number having the form $x + yi$. Accordingly, we can visually represent sums and products of complex numbers on a graph as well.

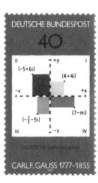

2. On the axes shown to the right, plot and label a point that correspond to each of the following complex numbers:

    a. $3 + 3i$
    b. $2 + i$
    c. $4 - 3i$
    d. $-2 + 2i$

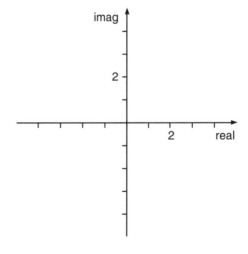

6.9B

Represent the complex numbers $3 + 1i$ and $1 + 2i$ as vectors v and w joining the origin $(0, 0)$ to the points $(3, 1)$ and $(1, 2)$ respectively. As shown to the right, the sum can be geometrically identified with the diagonal of a completed parallelogram.

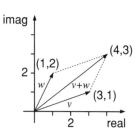

3. On the axes shown, form parallelograms with diagonals corresponding to these two sums computed in question 2.

   a. $(3 - 4i) + (1 + 1i)$
   b. $(-3 - 2i) + (1 + 4i)$

4. In the first column, expand each of the following binomial products into a single polynomial. In the second column, write the result after x is replaced by i. Reduce the expression using $i^2 = -1$ and record the result in the third column.

   Example: $(2 + 3x)(5 + 7x)$ $10 + 29x + 21x^2$ $10 + 29i + 21i^2$ $-11 + 29i$

   a. $(1 + 2x)(3 + 5x)$ _____   _____   _____
   b. $(3 - 4x)(5 - 7x)$ _____   _____   _____
   c. $(a + bx)(c + dx)$ _____   _____   _____

Multiplication of two complex numbers is defined by the formula $(a + bi)(c + di) = (ac - bd) + (ad + bc)i$. We now try to find a geometric representation using vectors. Identify the complex number $2 + 1i$ with vector v joining the origin $(0, 0)$ to the point $(2, 1)$. As shown, the vector w corresponds to the product $(2 + i)(2 + i) = (3 + 4i)$.

Notice that the angle β associated with vector w is double the angle α of vector v.

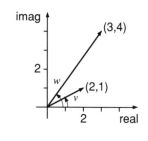

The size of a complex number $(a + bi)$ is $\sqrt{a^2 + b^2}$. This is the length of the vector representing the number.

5. Compute the length of vector v and w.

Notice that the length of w is the square of the length of v.

6.9C

6. On the axes to the right, draw vectors that correspond to each of the complex numbers given. Then draw a vector that represents the square of each complex number by doubling the angle and squaring the length. Label each pair of vectors with their lengths and angle sizes.

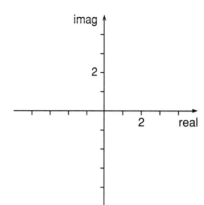

a. $(1 + i)$

b. $(-3/2 + (3/2)i)$

7. Compute the real and imaginary components of the vectors graphed in question 6 by applying the multiplication rule derived in question 4.

a. $(1 + x)(1 + x)$

b. $(-3/2 + (3/2)x)(-3/2 + (3/2)x)$

8. Compute the product of the following complex numbers and draw vectors that represent each of the corresponding numbers.

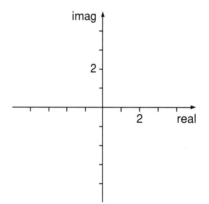

a. $(1/2 + i)\,(2 + i)$ _____

b. $(1 + i)\,(0 + 2i)$ _____

9. Determine the angles and sizes for the following complex numbers:

	$(1/2 + i)$	$(2 + i)$	$(0 + 5/2i)$	$(1 + i)$	$(0 + 2i)$	$(-2 + 2i)$
angle						
size						

10. Compare the results of 8 and 9. Write a conjecture that explains how a vector can be constructed geometrically to represent the product of two complex numbers. Express the angle of the product in terms of the angles of the factors. Express the size of the product in terms of the sizes of the factors.

6.10 ORBITS   6.10A

In Activities 6.4–6.6 we considered the nature of prisoner sets that are generated by functions of the form $y = x^2 + c$. For a given value of parameter c, some points remain prisoners while others escape under iteration.

In this activity, the value assigned to parameter c is allowed to be a complex number, and the main feature to be studied is once again the behavior of points under iteration. However, in this case we denote the variable by z to signify that the domain points are drawn from the complex plane. Each setting of parameter c defines a single function of the form $f(z) = z^2 + c$. For a given function, some points in the complex plane are prisoners while others escape to infinity under iteration.

Shown below are sample plots of two sequences of iterates generated by iterating an initial point through a function of the form $f(z) = z^2 + c$. Both sequences, or orbits, in these two examples spiral off to infinity.

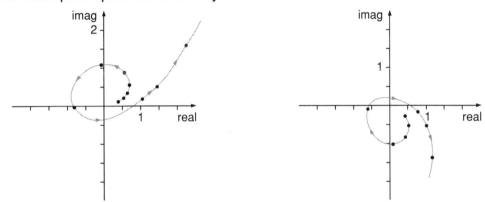

In the function $f(z) = z^2 + c$, let parameter $c = -1 + 0i$. On each graph of the complex plane shown below, a point (x, y) has been identified that corresponds to a complex number of the form $x + yi$. Use each point as an initial point for iteration through the function $f(z) = z^2 + (-1 + 0i)$. For each initial point, plot the points that correspond to the first five iterates. As on the sample graphs above, indicate each path with an arrow.

1.

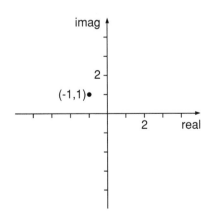

2.

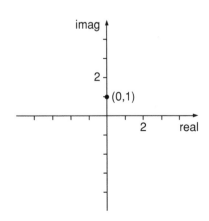

6.10B

3.

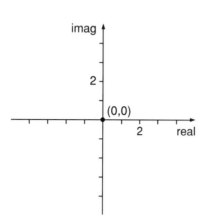

4.

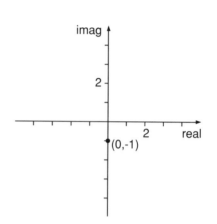

If you manually performed the complex number arithmetic as required by question 1, then you will appreciate the facility offered by a computer or programmable calculator. After you choose a value for the parameter $c = a + bi$, the programs shown below will iterate a complex number $x + yi$ of your choice through the function $f(z) = z^2 + c$.

5.   Enter the appropriate program into your graphing calculator.

Line	CASIO	Line	TEXAS INSTRUMENTS
1		1	:ClrHome
2		2	:Disp "A IN C = A+BI"
3	"A IN C=A+BI"? → A	3	:Input A
4		4	:Disp "B IN C = A+BI"
5	"B IN C=A+BI"? → B	5	:Input B
6	" "	6	:Disp " "
7		7	:Disp "X IN (X,Y)"
8	"X IN (X,Y)"? → X	8	:Input X
9		9	:Disp "Y IN (X,Y)"
10	"Y IN (X,Y)"? → Y	10	:Input Y
11	0→ N	11	:0→ N
12	Lbl 1	12	:Lbl 1
13	" "	13	:Disp " "
14	X → P	14	:X → P
15	Y → Q	15	:Y → Q
16	$P^2 - Q^2 + A → X$	16	:$P^2 - Q^2 + A → X$
17	$2PQ + B → Y$	17	:$2PQ + B → Y$
18	X △	18	:Disp X
19	Y △	19	:Disp Y
20		20	:Pause
21	N+1 → N	21	:N+1 → N
22	N<7 =>Goto 1	22	:If N<7
23		23	:Goto 1

Lines: 1–11   Initialize the counter N and allow for input of the parameter $c = (a, b)$ as well as the initial point (x, y).

10–23   Form the loop which generates the next iterate.

6.10C

In the function $f(z) = z^2 + c$, let parameter $c = 0.55 + 0.15i$. On each graph displayed below, a specific point (x, y) has been identified that corresponds to a complex number of the form $x + yi$. Use each of these points as an initial point for iteration through the function $f(z) = z^2 + (-1 - 1i)$. Then, for each initial point, plot the points that correspond to the first five iterates. Draw an arrow from each point to the next in the sequence.

6.

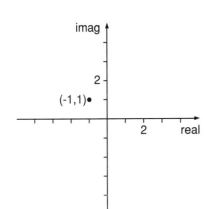

7.

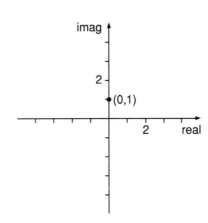

8.

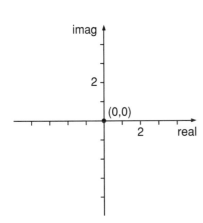

9.

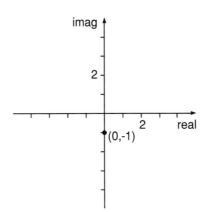

10. Select four complex numbers $x + yi$ of your own choosing. Iterate them through the function $f(z) = z^2 + (-1 - 1i)$ of questions 6–9. How many of your selected numbers correspond to prisoner points (x, y)? Make a conjecture as to why it is difficult to locate points that are in the prisoner set $f(z) = z^2 + (-1 - 1i)$.

As we will see in Activity 6.11, the set of all prisoner points for a given function $f(z) = z^2 + c$ will either be a connected set or a totally disconnected Cantor set. If the parameter c in the function $f(z) = z^2 + c$ happens to be selected so that the resulting prisoner set turns out to be disconnected, then it will be difficult to randomly select points for iteration that remain as prisoners. Such points are isolated from one another and form a dust-like Cantor set.

6.11 JULIA SETS

6.11A

As discussed in Activity 6.6 for functions $f(z) = z^2 + c$, if the parameter value c and the domain are both restricted to real numbers, then the collection of all prisoner points for a given function is either a connected invariant interval or a totally disconnected Cantor set of points. In the case of complex values for parameter c and a domain of complex points, the prisoner set is also either a connected region or a totally disconnected Cantor set of points.

Prisoner sets for two different values of parameter c in $f(z) = z^2 + c$ are displayed below.

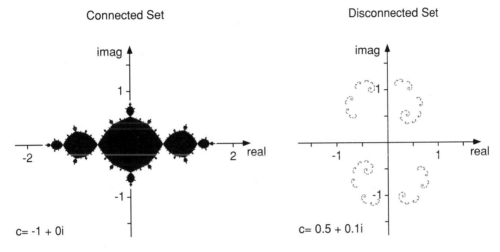

Connected Set   Disconnected Set

c= -1 + 0i   c= 0.5 + 0.1i

1. Let $f(z) = z^2 + (0 + 1i)$. Iterate $0 + 0i$ through the function f. Based upon the resulting pattern of iterates, explain why $0 + 0i$ is in the prisoner set of f.

2. Let $f(z) = z^2 + (0.5 + 0.1i)$. Show that the point $0 + 0i$ escapes to infinity under iteration through the function f.

For a given setting of parameter c, the behavior of the critical point $0 + 0i$ under iteration determines the nature of the prisoner set. When this critical point does not escape to infinity, the prisoner set is connected. When this critical point escapes to infinity, the prisoner set is a totally disconnected Cantor set.

3. What is the behavior of the point $0 + 0i$ when iterated through the function given by $f(z) = z^2 + (-1 + 0i)$? Is the prisoner set associated with $f(z) = z^2 + (-1 + 0i)$ connected or disconnected?

4. What is the behavior of $0 + 0i$ when iterated through the function $f(z) = z^2 + (-1 - 1i)$? Is the prisoner set connected or disconnected?

6.11B

For the function $f(z) = z^2 + c$, the points that lie on the boundary between the prisoner set and those points that escape to infinity are collectively referred to as the *Julia set*. When the prisoner set is connected, the Julia set consists of the boundary of this connected set. When the prisoner set is disconnected, the Julia set is the prisoner set itself.

5. In the function $f(z) = z^2 + c$, let $c = 0 + 0i$.

    a. Based upon the behavior of the critical point under iteration through $f(z) = z^2$, is the Julia set associated with this function $f(z)$ connected or disconnected?

    b. Arbitrarily select some points (a, b) inside the circle of radius $r = 1$ centered at the origin. Iterate the complex number $z = a + bi$ associated with each of your selected points through the function $f(z) = z^2$. For each iterate, determine if the associated point is closer to the origin or farther from the origin.

    c. Arbitrarily select some points (a, b) outside the circle of radius $r = 1$ centered at the origin. As in part (a), iterate the complex number $z = a + bi$ associated with each of your selected points through the function $f(z) = z^2$. For each iterate, determine if the associated point is closer to the circle or farther out towards infinity.

    d. Describe the Julia set defined by the function $f(z) = z^2$.

6. The Julia set for the specific function $f(z) = z^2 + (-1 + 0i)$ is shown below. Use your calculator to determine which of the following points are within the prisoner set of this function.

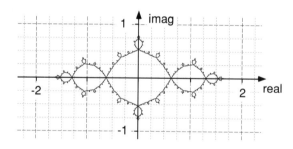

    a. (0,0)   b. (−1,0)   c. (0,0.7)   d. (0,0.8)

7. A *fixed point* is mapped to itself under a function or other transformation. Since fixed points obviously fail to escape to infinity, they are necessarily in the prisoner set. Can you identify any fixed points in the domain of $f(z) = z^2 + (-1 + 0i)$? Use your calculator and the programs provided in Activity 6.10 to check your results. Hint: Consider solving the equation $z = z^2 + (-1 + 0i)$.

6.12 THE MANDELBROT SET

6.12A

In Activities 6.5–6.6 we considered the nature of prisoner sets that are generated in the domain of real numbers by functions of the form $y = x^2 + c$ with real valued parameter c. By testing the single critical point $x = 0$, we determined if the prisoner set was an invariant interval or a totally disconnected set of points.

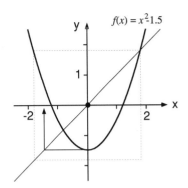

In Activity 6.11 we saw what happens with similar functions of the form $f(z) = z^2 + c$ but with complex values for parameter c and the complex plane as the domain. In this case again, the prisoner set is either a connected region or a totally disconnected set of points. Moreover, for a given setting of the parameter c, the behavior of the single critical point $0 + 0i$ under iteration determines which of these two cases occurs. When the critical point does not escape to infinity, the prisoner set is connected. When the critical point escapes to infinity, the prisoner set is totally disconnected.

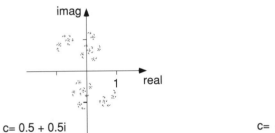
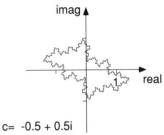

For a specific value of parameter c in $f(z) = z^2 + c$, the points that lie on the boundary between the prisoner set and those points that escape to infinity are collectively referred to as the Julia set.

1. For each of the given values of parameter c in $f(z) = z^2 + c$, use your calculator and the program on Activity sheet 6.12B to determine if the critical point under iteration through f escapes to infinity or fails to escape. If the critical point fails to escape and is therefore a prisoner, write P in the blank. If the critical point escapes, write E in the blank.

Parameter c	
$0.2 + 0.2i$	
$-0.3 + 0.2i$	
$-0.6 + 0.4i$	
$-1.3 + 0.1i$	
$-1.7 + 0.1i$	

Parameter c	
$0.2 - 0.2i$	
$-0.3 - 0.2i$	
$-0.6 - 0.4i$	
$-1.3 - 0.1i$	
$-1.7 - 0.1i$	

6.12B

The following programs perform iteration of the critical value $0 + 0i$ through the function of $f(z) = z^2 + c$. After each successive iteration, the coordinates a and b of the next point $a + bi$ will be displayed. If the associated outcome is more than 100 units from the origin, then the number N of iterations necessary to exceed this distance will be shown. Otherwise, after 30 iterations the word PRISONER is displayed.

Line	CASIO	Line	TEXAS INSTRUMENTS
1		1	:ClrHome
2		2	:Disp "A IN C = A+BI"
3	"A IN C=A+BI"? $\rightarrow$ A	3	:Input A
4		4	:Disp "B IN C = A+BI"
5	"B IN C=A+BI"? $\rightarrow$ B	5	:Input B
6	" "	6	:Disp " "
7	A $\rightarrow$ P	7	:A $\rightarrow$ P
8	B $\rightarrow$ Q	8	:B $\rightarrow$ Q
9	1 $\rightarrow$ N	9	:1 $\rightarrow$ N
10	Lbl 1	10	:Lbl 1
11	$P^2 - Q^2 + A \rightarrow R$	11	:$P^2 - Q^2 + A \rightarrow R$
12	2PQ + B $\rightarrow$ Q	12	:2PQ + B $\rightarrow$ Q
13	R $\triangle$	13	:Disp R
14		14	:Pause
15	Q $\triangle$	15	:Disp Q
16		16	:Pause
17	" "	17	:Disp " "
18	R $\rightarrow$ P	18	:R $\rightarrow$ P
19	N+1 $\rightarrow$ N	19	:N+1 $\rightarrow$ N
20	N>20 => Goto 2	20	:If N > 20
21		21	:Goto 2
22	$(R^2 + Q^2) < 10000$ => Goto 1	22	:If $(R^2 + Q^2) < 10000$
23		23	:Goto 1
24	" "	24	:Disp " "
25	"ITERATION COUNT"	25	:Disp "ITERATION COUNT"
26	N $\triangle$	26	:Disp N
27	Goto 3	27	:End
28	Lbl 2	28	:Lbl 2
29	" "	29	:Disp " "
30	"PRISONER"	30	:Disp "PRISONER"
31	Lbl 3	31	:

Lines: 1–9   Initialize the counter N and allow for input of parameter c.

10–23   Form the loop which counts the number of iterations required before the values of the successive iterates exceed 100.

24–31   Display the resulting iteration count N or the word PRISONER.

6.12C

2. Each complex number $c = a + bi$ listed in question 1 corresponds to a point (a, b) within the left-hand grid shown below. For each point (a, b), find the cell within the grid that contains it. Write either a P or an E in the cell according to whether the critical point is a prisoner or one that escapes.

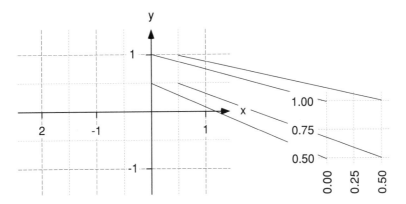

3. Select the center point from within each cell that has not yet been labeled. Each of your selected points (a, b) determines a function of the form $y = z^2 + c$ with $c = a + bi$. Use the calculator programs to determine if the critical point under iteration through f escapes to infinity or fails to escape. Label the respective cells with P or E according to your results.

4. If we partition each cell in the above left-hand grid into subsections, then a refined identification can be implemented by selecting values for $c = a + bi$ from within each subsection. For example, the right-hand grid in the graph above displays four smaller subsections. Select the center point from within each of these four cells. Use your calculator and one of the programs on the previous page to determine if the critical point is a prisoner or one that escapes. As before, label each of the four cells with P or E according to your results.

The *Mandelbrot set* identifies those parameter values $c = a + bi$ in the function $f(z) = z^2 + c$ for which iteration of the critical point $z = 0 + 0i$ yields a sequence that fails to escape to infinity. Such parameter values are represented visually on the map by black points, and they are associated with connected Julia sets. Points outside of the Mandelbrot set relate to parameter values for which the associated functions iterate the critical point $z = 0 + 0i$ to infinity. These points correspond to disconnected Julia sets.

5. Color each cell in the above grid black that contains a P.

6.12D

The process initiated in question 3 that labels points within a refined grid can be continued for even finer grids.

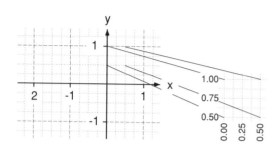

Coloring each cell black that contains a P in a very fine grid gives a very good representation of those points that form the Mandelbrot set. Such a representation of the Mandelbrot set is shown at the right side.

Points outside of the Mandelbrot set relate to parameter values for which the associated functions iterate the critical point $z = 0 + 0i$ to infinity. Such points correspond to disconnected Julia sets. Each point within the Mandelbrot set relates to a parameter value for which the critical point $z = 0 + 0i$ is a prisoner under iteration through the associated function. Such points correspond to connected Julia sets.

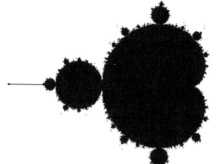

It is known that the Mandelbrot set is a connected set with very fine dendrites at its boundary. The method that we described for obtaining a representation of the Mandelbrot set cannot show these features. Rather, you see some isolated points close to a big connected part. Maybe the grid used is not fine enough? Unfortunately, changing the grid size does not help.

A much more sophisticated algorithm is required to reveal the full beauty of the Mandelbrot set. Such an algorithm was used to compute the second, improved representation shown here. When comparing the two images, you can see how the fine dendrites match the isolated points of the first picture.

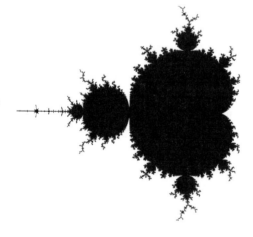

6.13 COLORING THE MANDELBROT SET   6.13A

Points (a, b) outside of the Mandelbrot set correspond to parameter values $c = a + bi$ in the function $f(z) = z^2 + c$ for which iteration of the critical point $z = 0 + 0i$ yields a sequence of iterates that escapes to infinity. Colors can be assigned to these points according to the rate of escape. We measure this rate by reference to the number of iterations N required before the iterates lie more than 100 units from the origin.

1. For each of the given values of parameter c in $f(z) = z^2 + c$, use your calculator and the programs supplied with Activity 6.12 to determine the number N of iterations required before the critical point under iteration through f is more than 100 units from the origin. Record this number N in the chart below. If the critical point is a prisoner and fails to escape, then write P in the blank.

Parameter c	Iterations N
$0.2 + 0.2i$	
$-0.3 + 0.2i$	
$-0.6 + 0.4i$	
$-1.3 + 0.1i$	
$-1.7 + 0.1i$	

Parameter c	Iterations N
$0.2 - 0.2i$	
$-0.3 - 0.2i$	
$-0.6 - 0.4i$	
$-1.3 - 0.1i$	
$-1.7 - 0.1i$	

2. Each complex number $c = a + bi$ listed in question 1 corresponds to a point (a, b) within the left-hand grid displayed below. For each point (a, b), find the cell that contains it in the grid and use colored markers to shade the cell according to the following scheme.

    Black   if the sequence fails to escape
    Yellow   if the sequence exceeds 100 on iteration 1 through 4
    Orange if the sequence exceeds 100 on iteration 5 through 8
    Red   if the sequence exceeds 100 on iteration 9 through 12
    Green   if the sequence exceeds 100 on iteration 13 through 16
    Purple if the sequence exceeds 100 on iteration 17 and higher

3. For each cell that has not yet been colored select the point at the center within. Each of these points (a, b) determines a function of the form $f(z) = z^2 + c$ with $c = a + bi$. Use the calculator programs to determine if the critical point under iteration through f escapes to infinity or fails to escape. Apply color to those cells according to the iteration count N as before.

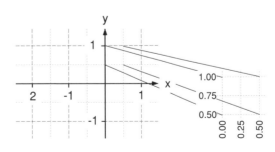

4. If we partition each cell in the above left-hand grid into subsections, then a refined coloring process can be implemented by selecting values for $c = a + bi$ from within each subsection. For example, the right-hand grid in the graph displays four subsections. Select the center point from within each of these four cells. Use the calculator programs to determine the number N of iterations required before the critical point iterates to more than 100 units from the origin. Color the four cells according to your results.

The *Mandelbrot set* identifies those parameter values $c = a + bi$ in the function $f(z) = z^2 + c$ for which iteration of the critical point $z = 0 + 0i$ yields a sequence that fails to escape to infinity. Such parameter values are represented visually on the map by black points, and they are associated with connected Julia sets. Points outside of the Mandelbrot set relate to parameter values for which the associated functions iterate the critical point $z = 0 + 0i$ to infinity. These points correspond to disconnected Julia sets, and they are colored according to the rate at which the critical point escapes. The exceedingly irregular boundary of the set of black points forms a barrier between these two behaviors. Magnified portions of this boundary are displayed below.

 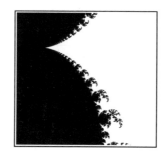 

Upon reviewing the activities in Unit 6, you might notice that selecting real values for parameter c in both the real valued functions $f(x) = x^2 + c$ as well as in the complex valued functions $f(z) = z^2 + c$ induces a correspondence between the basic features of the resulting prisoner sets. The chart on the next page indicates the nature of this correspondence.

5. Using the chart on the following page, compare the prisoner sets for the above classes of real and complex functions when parameter c is chosen to be a real number. In what ways will the prisoner sets be alike for a given value of c within the interval $(-\infty, -2)$? Within the interval $[-2, 1/4]$? Why is this comparison made difficult for values of $c > 1/4$?

6.13C

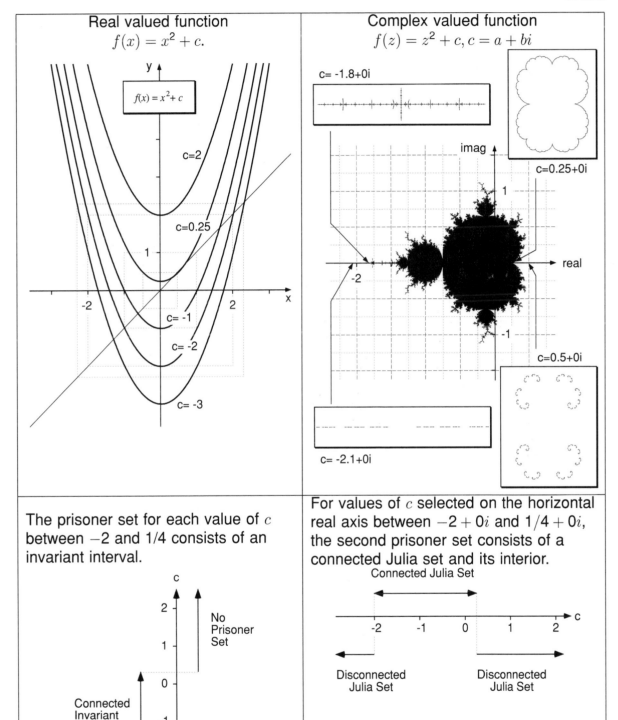

Real valued function
$$f(x) = x^2 + c.$$

$f(x) = x^2 + c$

c=2

c=0.25

c= -1

c= -2

c= -3

Complex valued function
$$f(z) = z^2 + c, \, c = a + bi$$

c= -1.8+0i

c=0.25+0i

imag

1

real

-2

-1

c=0.5+0i

c= -2.1+0i

The prisoner set for each value of c between -2 and $1/4$ consists of an invariant interval.

c

2

1

0

-1

-2

No Prisoner Set

Connected Invariant Interval

Cantor Set

For values of c selected on the horizontal real axis between $-2 + 0i$ and $1/4 + 0i$, the second prisoner set consists of a connected Julia set and its interior.

Connected Julia Set

-2   -1   0   1   2   c

Disconnected Julia Set

Disconnected Julia Set

6.13D

Programming the Mandelbrot set and enlargements

The Mandelbrot set is used — or misused — worldwide to drive computers crazy. Home Computers and PC'c do it, workstations do it, and even Super Computers are regularly pushed to their limits with the Mandelbrot set. Clearly, programmable, hand-held pocket calculators do not stand a chance competing against their big brothers. Given their limitations it is astonishing that they are at all capable of rendering images of the Mandelbrot set. Furthermore, it is amazing how much they can achieve with so little programming effort. On the right is a list of the complete TI-81 code consisting of mere 48 lines. Yet it provides the main features that one expects (except for color display).

- Computation of a global view of the set.
- Interactive positioning of a zoom window.
- Computation and display of the selected enlargement.

The last two steps may be repeated. Here are the operating instructions:

1. Type in the program and run it.
2. Type "0" when prompted "ZOOM? (0,1)".
3. Enter the number N of maximal iterations upon "MAX ITER?"
4. Now the program starts to compute (this may take a while).
5. After termination of the program a picture of the Mandelbrot set (or a section of it) is on the display. Now, a window for an enlargement can be chosen. First, push the ZOOM button.
6. Select the item "1:Box" from the displayed zoom menu.
7. Using the four cursor control buttons position the cursor to the upper left corner of the desired zoom window. The coordinates of the point are displayed. Push ENTER.
8. Repeat for the lower right hand corner. The outline of the window is dragged across the display. Push ENTER and CLEAR.
9. Start the program again. This time, enter "1" when responding to "ZOOM? (0,1)". Continue with step 3.

Explanations: When setting $N = 15$ the overview picture takes 25 minutes to compute. Enlargements require longer times. Setting the zoom box in steps 6–8 defines the variables Xmin to Ymax which, when not zooming, are defined in lines 9–17 of the program. The display consists of 64 lines of 96 pixels each. When computing the default overview the symmetry of the Mandelbrot set about the real axis is exploited and only 32 lines are computed. The program variables J and I denote the line and pixel number. K is the iteration counter. The loops in the program structure are indicated by the arrows. A similar program can be written for the CASIO graphing calculator.

```
1  All-off
2  Clr-Draw
3  Disp "ZOOM? (0,1)"
4  Input Q
5  Disp "MAX ITER?"
6  Input N
7  63 → Z
8  If Q
9  Goto 0
10 │   -1.2 → Ymin
11 │    1.2 → Ymax
12 │    0.5 → Yscl
13 │   -2.6 → Xmin
14 │    1.0 → Xmax
15 │    0.5 → Xscl
16 ▼   31  → Z
17 Lbl 0
18 DispGraph
19 (Xmax - Xmin) / 95 → C
20 (Ymax - Ymin) / 63 → D
21 0 → J
22 Lbl J
23 ▲  Ymin + JD → B
24 │  0 → I
25 │  Lbl I
26 │     ▲  Xmin + IC → A
27 │     │  A → X
28 │     │  B → Y
29 │     │  1 → K
30 │     │  Lbl 1
31 │     │     ▲  XX → U
32 │     │     │  YY → V
33 │     │     │  2XY + B → Y
34 │     │     │  U - V + A → X
35 │     │     │  K + 1 → K
36 │     │     │  If U + V > 100
37 │     │     │  Goto 2 ──┐
38 │     │     │  If K < N │
39 │     │  Goto 1         │
40 │     │  Pnt-On (A,B)   │
41 │     │  If Z=31        │
42 │     │  Pnt-On (A,-B)  │
43 │     │  Lbl 2 ◄────────┘
44 │     │  IS > (I,95)
45 │  Goto I
46 │  IS > (J,Z)
47 Goto J
48 End
```

Answers

UNIT 4   Iteration

ACTIVITY 4.1

1. a. Steps in toward a fixed point
    b. Steps out away from a fixed point
2. a. The intersection point is $(-1.3, -1.3)$; an attractor
    b. The intersection point is $(-1.3, -1.3)$; a repeller
3. A repeller
4. An attractor

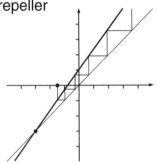

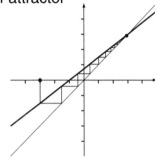

5. a. Spirals in toward a fixed point
    b. Spirals out away from a fixed point
6. a. The intersection point is $(1, 1)$; an attractor
    b. The intersection point is $(0, 0)$; a repeller
7. An attractor
8. A repeller

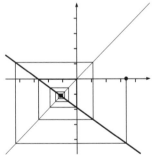

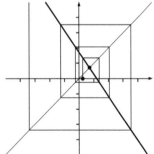

9. Staircase; repeller
10. Spiral; repeller
11. Spiral; attractor
12. Staircase; attractor
13. 1/4; 1/2; 1; 2
14. 1; -1/2; 1/4; −1/8
15. −1/4; 1/2; −1; 2
16. −2; −1; −1/2; −1/4
17. The starting point has basically no effect on the iteration pattern as long as it is not the x-coordinate of the fixed point.
18. If the slope of the straight line is positive, then the iteration pattern is that of a staircase. If the slope of the straight line is negative, then the iteration pattern is a spiral. More specifically, for $m < -1$, repel and spiral out; for $-1 < m < 0$, attract and spiral in; for $0 < m < 1$, attract and staircase in; and for $1 < m$, repel and staircase out.

19. It goes from a spiral out to a spiral in pattern.
20. It switches from a spiral in pattern to a staircase in pattern.
21. It goes from a staircase in to a staircase out pattern.
22. If the slope of the line is 1 and it has a non-zero y-intercept, it will staircase to the left and down if below the fixed line; and staircase to the right and up if above the fixed line. If the lines y-intercept is zero and its slope is 1, then each starting point is a fixed point attractor without an iteration pattern.
23. If the slope is -1, then the iteration pattern is always a clockwise box.

ACTIVITY 4.2

1. 16; 17   2. -6; 36   3. 400; 375   4. 0; 1
5. 16   6. 4   7. 1   8. 4; 4; -1
9. 5; 1; 11   10. 1; 5; -4 11. 4; 4; -1
12. a. $a^2 + 1$ b. $a^2 + 2a + 1$ c. a^4 d. $a + 2$
13. 0.447, 0.669, 0.818, 0.904, 0.951, 0.975, 0.988, 0.994
14. 0.922, 0.960, 0.980, 0.990, 0.995, 0.997, 0.999, 0.999
15. 8, 2.828, 1.682, 1.297, 1.139, 1.067, 1.033, 1.016
16. 43.012 6.558, 2.561, 1.600, 1.265, 1.125, 1.061, 1.030
17. They all appear to be approaching 1, but from different directions.
18. a. 1, 1, 1, 1, 1
    b. 25, 625, 390625, $1.526 \cdot 10^{11}$, $2.328 \cdot 10^{22}$
    c. 0.04, 0.0016, $2.56 \cdot 10^{-6}$, $6.55 \cdot 10^{-12}$, $42.9 \cdot 10^{-24}$
    d. 9, 81, 6561, 43046721, $1.853 \cdot 10^{15}$
19. The iterates get large quickly and approach $+\infty$.
20. The iterates get small quickly and approach zero.
21. If initial values are in the interval $-1 < x_0 < 0$, they will iterate toward zero.
22. If the initial values are less than -1, they will iterate quickly toward $+\infty$.
23. $f(x) = x + 1$
24. $f(x) = 2x$
25. $f(x) = -3x$
26. $f(x) = x + 5$
27. $f(x) = .5x$
28. Arithmetic: 23 and 26; Geometric: 24, 25, and 27

ACTIVITY 4.3

1.

2.

3.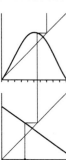

4.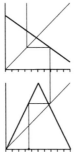

5.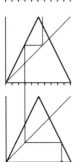

6.

4.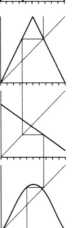

5.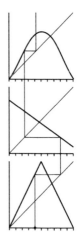

6.

7.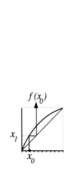

8.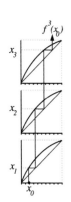

9. $x_1 = f(x_0)$, $x_2 = f^2(x_0)$, $x_3 = f^3(x_0)$, $x_4 = f^4(x_0)$

10.

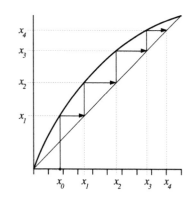

ACTIVITY 4.4

1.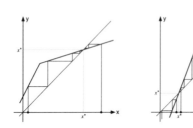

2.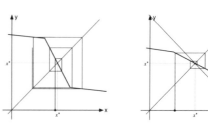

3.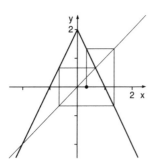

4.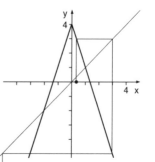

The graphical iteration moves about the interval $[-2, 2]$ in an irregular way.

The graphical iteration tends to negative infinity.

5. $1/3 \to 4/3 \to -2/3 \to 2/3 \to 2/3 \to 2/3 \to 2/3$.
     After 3 iterations the fixed point at $x = 2/3$ is reached.

6. $1/3 \to 3 \to -5 \to -11 \to -29 \to -83 \to -245$.
     The iteration tends to negative infinity.

7.   For $f(x) = -2|x| + 2$ the graphical iteration does not yield the same result as the exact iteration because of small errors in the process which move the iteration sequence away from the repelling fixed point at $x = 2/3$. For $f(x) = -3|x| + 4$ the qualitative behavior in graphical and exact iteration is the same.

ACTIVITY 4.5

1.   a. $y > 3$ b. $y = 3$ c. $y < 3$
2.   a. 3   b. 3   c. 5   d. 0
3. ± 1
4.

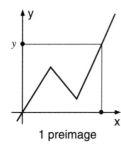
1 preimage

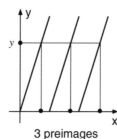
3 preimages

3 preimages

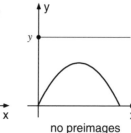
no preimages

5.   6.   7.

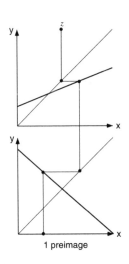

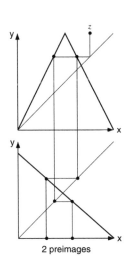

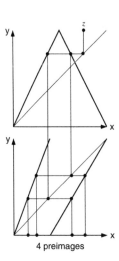

1 preimage   2 preimages   4 preimages

ACTIVITY 4.6

1. The interval (b, c) staircases up and to the right. After each iteration, the interval gets larger. In fact, after each iteration, it actually doubles in width.
2. The interval is expanding.
3. Point b: $0.19 \rightarrow 0.38 \rightarrow 0.76 \rightarrow 1.52 \rightarrow 3.04 \rightarrow 6.08$
    Point c: $0.21 \rightarrow 0.42 \rightarrow 0.84 \rightarrow 1.68 \rightarrow 3.36 \rightarrow 6.72$
4. Interval width: $0.02 \rightarrow 0.04 \rightarrow 0.08 \rightarrow 0.16 \rightarrow 0.32 \rightarrow 0.64$
    Absolute error: $0.01 \rightarrow 0.02 \rightarrow 0.04 \rightarrow 0.08 \rightarrow 0.16 \rightarrow 0.32$
5. With each iteration the absolute error doubles.
6. The relative errors at stages 3, 4, and 5 are 5%, 5%, and 5%.; The relative error remains constant.
7. $0.02 \cdot 2^n$; $0.02 \cdot 2^{n-1}$; 5%
8. The error interval is spiraling in clockwise to the origin and it is being compressed.
9. Interval width: 0.00125; absolute error: 0.000625; relative error: 2%
10. As the number of iterations increases, the absolute error is compressed or reduced by a factor of two after each one. On the other hand, as the number of iterations increases, the relative error remains constant at 2%.
11. Interval width compresses; absolute error decreases; relative error remains constant
12. Interval width expands; absolute error increases; relative error remains constant
13. Interval width expands; absolute error increases; relative error remains constant
14. Interval width remains constant; absolute error remains constant; relative error decreases

ACTIVITY 4.7

1. The graphical iteration moves in staircase fashion, stepping in toward the origin that serves as an attractor.
2. The graphical iteration and iteration sequences for both $x_0 = 0.7$ and $x_0 = 0.3$ are identical except for the first step and initial values.

3. yes
4. compression
5. staircase
6. (3/8, 3/8) is the other fixed point and is an attractor.
7. Compression at (3/8,3/8). Expansion at (0,0).
8.

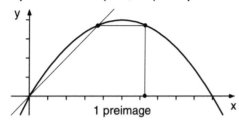

1 preimage

9. Student answers will vary somewhat depending on the care with which they read
   the graphical iteration. However, their answers should approximate the following:
   0.10, 0.25, 0.53, 0.70, 0.59, 0.68, 0.61, 0.67. The iteration behavior starts
   staircasing up and to the right, but after two iterations switches to spiraling in on
   an attractor.
10. (9/14, 9/14); an attractor; interval compression
11. Initial points will vary, but the iteration behaviors will all typically be the same as
    for questions 9 and 10.
12.

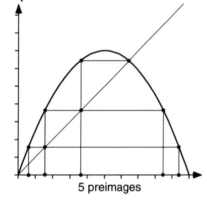

5 preimages

13. The initial value $x_0 = 0$, repeatedly iterates to itself. The initial value $x_0 = 1$
    iterates immediately to 0 and all subsequent iterates are also 0. Speculations will
    vary, however all the points outside the closed interval [0, 1] will iterate toward
    $-\infty$.
14. 0.10, 0.29, 0.66, 0.72, 0.64, 0.73, 0.62, 0.75
15. They are different, the iterates in question 9 were slowly converging to a single
    value; here the iterates converge to two values which appear fixed, alternating
    back and forth between them.
16. The iteration behavior was first a staircase, quickly switching to a slow spiral,
    ending up in a box pattern.; No, the attractor is a periodic cycle. Both interval
    expansion and compression take place around the box pattern, each cancelling
    the other out in terms of their net effect.

17.

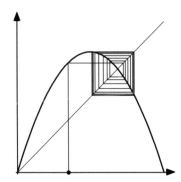

18. yes; yes; spiraling out; Both interval expansion and compression take place around the box pattern, each cancelling the other out in terms of their net affect.
19. They are identical except for their initial values.
20. 0.6875 is a fixed point so the graphical iteration is a single vertical line segment.
21. The initial value $x_0 = 1$ immediately iterates to 0 and then stays fixed at 0 for all subsequent iterates.
22. They all iterate off toward $-\infty$.
23. They all iterate off toward $-\infty$.
24. Besides the repelling fixed point itself, there is one initial value that iterates to the fixed point in only one iteration; there are two values that do this in two, and another two values that do it in three iterations.

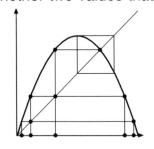   25.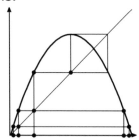

ACTIVITY 4.8

1.

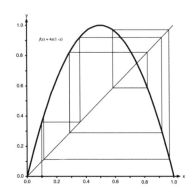

2. Answers will vary widely. A possible set close to the exact solution is: 0.10, 0.36, 0.92, 0.29, 0.82, 0.58, 0.97, 0.11.

3. A mixture of staircasing and spiraling.; It does not appear to have either an attractor or a repeller.; No; No pattern is apparent, it seems to iterate all over the closed interval [0, 1].

4. Even with the same choice of initial values, again the answers will vary widely.

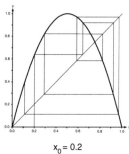

$x_0 = 0.2$

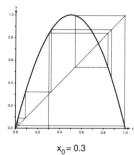

$x_0 = 0.3$

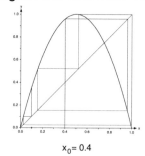

$x_0 = 0.4$

5. more chaotic
6. The sequence of iterates appears to be: 0.5, 1, 0, 0, ...
7. The sequence of iterates appears to be: 0.75, 0.75, 0.75, ... (a single vertical segment).
8. The sequence of iterates appears to be: 1, 0, 0, 0, ...
9. The sequence of iterates appears to be: 0.25, 0.75, 0.75, 0.75, ...

ACTIVITY 4.9

1. Except for $a = 4$, where sensitivity is present, the results are the same as shown in the graphs.

2.
$a =$	1.50	2.90	3.24	3.90
Chaos				•
Period-2 attractor			•	
Fixed point, spiral in		•		
Fixed point, staircase in	•			

3. Results are independent of the initial values. Period-4 cycle: 0.501, 0.875, 0.383, 0.827.

4.
$a =$	2.95	3.05	3.50	3.68	3.74	3.80	3.84
Chaos				•		•	
Period-5 attractor					•		
Period-4 attractor			•				
Period-3 attractor							•
Period-2 attractor		•					
Fixed point	•						

ACTIVITY 4.10

1. $x_0 = 2.5, x_1 = 1.58, x_2 = 1.26, x_3 = 1.12$;
   $x_0 = 3.0, x_1 = 1.73, x_2 = 1.32, x_3 = 1.15$.

2. $x_0 = 1.1, x_1 = 1.21, x_2 = 1.46, x_3 = 2.14$;
    $x_0 = 1.2, x_1 = 1.44, x_2 = 2.07, x_3 = 4.30$.

3. Differences for question 1: 0.5, 0.15, 0.06, 0.03 (compression).
    Differences for question 2: 0.1, 0.23, 0.61, 2.26 (expansion).

4.    5.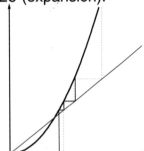

6. For $0 < x < 0.375$ and $0.625 < x < 1$.

7. $0.375 < x < 0.625$.

8. If the absolute value of the slope of a tangent to a function at x_i is greater than 1, then one step of graphical iteration will expand a very small interval from a to b when $a < x_i < b$; otherwise we have interval compression when the absolute of the slope of the tangent is less than 1.

9. It staircases in toward a fixed attractor point.

10. It staircases out away from a fixed repeller point.

11. If graphical iteration compresses an interval while staircasing, then it is converging on an attractor point; otherwise, expansion occurs while staircasing away from a repeller point.

12. Graphical iteration compresses small intervals in the interval $0.25 < x < 0.75$.

13. Graphical iteration compresses small intervals in the interval $0.375 < x < 0.625$.

14. Graphical iteration compresses small intervals in the interval
    $0.34375 < x < 0.65625$.

15. Graphical iteration compresses small intervals in the interval $0 < x < 1$.

16. Error expansion is most likely to occur for the logistics function $f(x) = ax(1 - x)$ when $a > 2$ and increases dramatically as a nears the parameter value 4.

ACTIVITY 4.11

1.

x_0	x_1	x_2	x_3	x_4	x_5	x_6	x_7
0.100	0.252	0.528	0.698	0.590	0.677	0.612	0.665
0.450	0.693	0.596	0.674	0.615	0.663	0.626	0.656
0.700	0.588	0.678	0.611	0.666	0.623	0.657	0.631

Subsequent iterations would continue to converge on 0.6429.

2.

x_0	x_1	x_2	x_3	x_4	x_5	x_6	x_7
0.100	0.288	0.656	0.722	0.642	0.735	0.623	0.752
0.450	0.792	0.527	0.798	0.517	0.799	0.514	0.799
0.700	0.672	0.705	0.665	0.713	0.655	0.723	0.641

3. Visually and numerically the results confirm the presence of a period one and a period two attractor.

4.

x_0	x_1	x_2	x_3	x_4	x_5	x_6	x_7
0.100	0.360	0.922	0.289	0.822	0.585	0.971	0.113
0.450	0.990	0.040	0.152	0.516	0.999	0.004	0.016
0.700	0.840	0.538	0.994	0.023	0.088	0.321	0.872

5. Visually and numerically the iteration pattern moves all about within [0,1].

6.

a	x_0	x_1	x_2	x_3	x_4	x_5	x_6	x_7
2.8	0.100	0.252	0.527	0.697	0.591	0.676	0.613	0.664
3.2	0.100	0.288	0.656	0.722	0.642	0.735	0.623	0.751
4.0	0.100	0.360	0.921	0.291	0.825	0.577	0.976	0.093

7. For $a = 2.8$ and $a = 3.2$ the differences are very minor. Only the last digit is off by at most 1 unit corresonding to about 0.11% and 0.07% in x_7. However, for $a = 4$, these small deviations are growing with the number of iterations. At the seventh iteration the difference is already about 0.02 or 18%.

9.

Iterations	10 decimal places	3 decimal places	difference
0	0.1000000000	0.100	0.000
6	0.9708133262	0.976	0.005
7	0.1133392474	0.093	0.020
8	0.4019738495	0.337	0.065
9	0.9615634952	0.893	0.069
10	0.1478365595	0.382	0.234
11	0.5039236447	0.944	0.440
12	0.9999384200	0.211	0.789

10. The numerical results are in the table above (right column). They indicate that the low precision computations do not approximate the more precise computations (10 digits) already from the tenth iteration on. From then on the errors are of the same order as the correct numbers.

11.

Iterations	10 decimal places	10 decimal places	difference
0	0.1000000000	0.1000000001	0.000
10	0.1478365595	0.1478364473	0.000
20	0.8200138632	0.8201379178	0.000
25	0.9863790703	0.9851552298	0.001
30	0.3203298063	0.4815763382	0.162

Even at very high precision calculations, a small initial error magnifies and after some 30 iterations the deviation already half as large as the result itself, thus implying errors of the order of 100% for the following iterations.

ACTIVITY 4.12

1. The origin or $x = 0$.
2. Yes, they all iterate to the attractor at $x = 0$.; Points in the intervals $-1 < x_0 < 0$ and $0 \le x_0 \le 1$ also iterate to $x = 0$.; the point -1 is an unstable fixed point (repeller), 0 is a stable fixed point (attractor) and 1 iterates to the attractor 0 in one iteration and then stays there.
3. The long term behavior of a point in the interval $0 \le x_0 \le 1$ ultimately iterates to the left toward negative infinity.
4. Two such points exist, 0 and 0.8. Both are unstable fixed points (repellers). Also their preimages do not iterate to infinity.
5. Yes.
6. $-(0.5 + \sqrt{0.9}) \le x_0 \le 0.5 + \sqrt{0.9} \approx 1.449$.
7. It spirals out to a box or rectangular graphical iteration pattern which indicates the presence of a period two attractor, namely 0 and -1.
8. $-(0.5 + \sqrt{1.25}) \le x_0 \le 0.5 + \sqrt{1.25} \approx 1.618$.
9. The iteration pattern is staircasing to the right and up toward positive infinity. No, there are no repellers or attractors.
10. At first it would compress for a few iterations and then suddenly begin to expand.
11. For $c = 0.40, 0.35, 0.30$, it would compress at first for a few iterations and then suddenly begin to expand. For $c = 0.25$, the iteration sequence continues to compress and appears to be converging onto an attractor.
12. For $c = -0.6, -0.7$, the iteration pattern is spiraling clockwise in onto a fixed point. For $c = -0.8, -0.9$, the iteration pattern is spiraling clockwise out onto a period two attractor.
13. For values of $c > -2$, all the iterates appear to stay in the interval $-2 \le x_0 \le 2$. But for values of $c < -2$, any point iterates off toward positive infinity.

ACTIVITY 4.13

1.

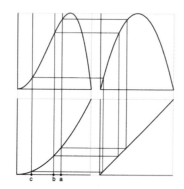

2. $g(f(x)) = 4x^2(1 - x^2) = 4x^2 - 4x^4$
3. 0.25 and 0.75; yes
4. 0.04 and 0.1536
5.

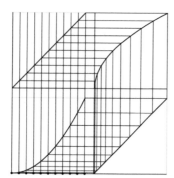

6. $f(x)$ and $g(x)$ are inverse functions and thus in composition with each other, produce the identity function, i.e. $g(f(x)) = f(g(x)) = x$.

7.   8.

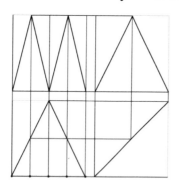   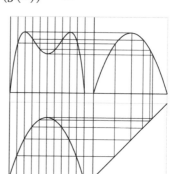

9. There are four points of intersection.; 0, 0.51, 0.68, and 0.799; yes
10. See student work on their graphing calculator.
11. See student work.

UNIT 5   Chaos

ACTIVITY 5.1

1. Yes
2. Yes
2. Yes
3.

4.

5. Yes
6.

7.

8. Yes
9.

10.

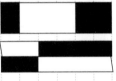

11.

12.

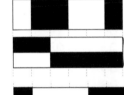

13.   Stretch-and-fold

L	K	J	I	H	G
A	B	C	D	E	F

Stretch-and-fold

F		E		D	
G		H		I	
L		K		J	
A		B		C	

14.   Stretch-and-cut-and-paste

G	H	I	J	K	L
A	B	C	D	E	F

Stretch-and-fold

F		E		D	
L		K		J	
G		H		I	
A		B		C	

15. Same letters in columns; order within column different; yes
16.–17. 14/25
18.–19. 14/25
20. Yes
22. 7/25 → 14/25 → 22/25 → 6/25 → 12/25 → 24/25
      8/25 → 16/25 → 7/25 → 14/25 → 22/25 → 6/25
23. 7/25 → 14/25 → 3/25 → 6/25 → 12/25 → 24/25
      8/25 → 16/25 → 7/25 → 14/25 → 3/25 → 6/25
24. 5/25 → 10/25 → 20/25 → 10/25 → 20/25 → 10/25

25. $5/25 \rightarrow 10/25 \rightarrow 20/25 \rightarrow 15/25 \rightarrow 5/25 \rightarrow 10/25$

ACTIVITY 5.2

1.–2. 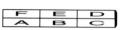   3.–4.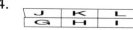

5. 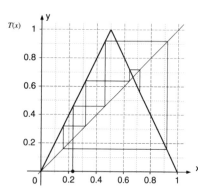   6.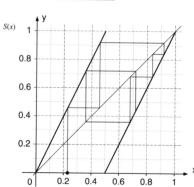

7. $0.23 \rightarrow 0.46 \rightarrow 0.92 \rightarrow 0.16 \rightarrow 0.32 \rightarrow 0.64 \rightarrow 0.72$
8. $0.23 \rightarrow 0.46 \rightarrow 0.92 \rightarrow 0.84 \rightarrow 0.68 \rightarrow 0.36 \rightarrow 0.72$
9. $0.22 \rightarrow 0.44 \rightarrow 0.88 \rightarrow 0.24 \rightarrow 0.48 \rightarrow 0.96 \rightarrow 0.08$
     $0.24 \rightarrow 0.48 \rightarrow 0.96 \rightarrow 0.08 \rightarrow 0.16 \rightarrow 0.32 \rightarrow 0.64$
10. $0.22 \rightarrow 0.44 \rightarrow 0.88 \rightarrow 0.76 \rightarrow 0.52 \rightarrow 0.04 \rightarrow 0.08$
     $0.24 \rightarrow 0.48 \rightarrow 0.96 \rightarrow 0.92 \rightarrow 0.84 \rightarrow 0.68 \rightarrow 0.36$

11. 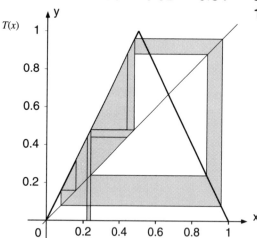   12.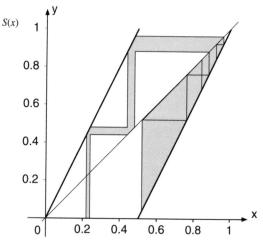

13. After 6 (tent function) resp. 5 (saw tooth function) iterations
14. Fixed; grain remains at its position
15. $0.03 \times 2^3 = 0.24$
16. $2/5 \rightarrow 4/5 \rightarrow 2/5 \rightarrow 4/5 \rightarrow 2/5$
17. $1/3 \rightarrow 2/3 \rightarrow 1/3 \rightarrow 2/3 \rightarrow 1/3$
18. $1/7 \rightarrow 2/7 \rightarrow 4/7 \rightarrow 1/7$

ACTIVITY 5.3

1. $3^4 = 1010001_{two}$; $3^5 = 11110011_{two}$
2. $(3/4)^4 = 0.01010001_{two}$; $(3/4)^5 = 0.0011110011_{two}$
6. 6/25
7. 1/4
8. 2/3
9.

Second stage

	0.00...		0.01...		0.10...		0.11...	
0				1/2				1

Third stage

	0.000...	0.001...	0.010...	0.011...	0.100...	0.101...	0.110...	0.111...	
0					1/2				1

10. The first three
11. 16, 32, 64; length is 1/16, 1/32, 1/64

12.

input a	fract(a)	int(a)
4/7	4/7	0
40/7	5/7	5
50/7	1/7	7
10/7	3/7	1
30/7	2/7	4
20/7	6/7	2
60/7	4/7	8

13. $0.\overline{857142}$
14. $0.\overline{27}$
15. $0.4\overline{615}$

17.

input a	fract(a)	int(a)
$1.8\overline{666}$	$0.8\overline{666}$	1
$1.7\overline{333}$	$0.7\overline{333}$	1
$1.4\overline{666}$	$0.4\overline{666}$	1
$0.9\overline{333}$	$0.9\overline{333}$	0
$1.8\overline{666}$	$0.8\overline{666}$	1

$1.8\overline{666}_{ten} = 0.\overline{1101}_{two}$

input a	fract(a)	int(a)
4/7	4/7	0
8/7	1/7	1
2/7	2/7	0
4/7	4/7	0
8/7	1/7	1

$4/7 = 0.1\overline{001}$

18. $0.\overline{1001}$
19. $0.\overline{0111000}$
20. $0.\overline{001001110110}$

ACTIVITY 5.4

1. $x_2 = 0.11\overline{01}$, $x_3 = 0.1\overline{01}$, $x_4 = 0.01\overline{01}$, $x_5 = 0.\overline{01}$
2. $0.011 \rightarrow 0.11 \rightarrow 0.1 \rightarrow 0.0$
3. $0.01\overline{01} \rightarrow 0.1\overline{01} \rightarrow 0.01\overline{01} \rightarrow 0.1\overline{01}$

4. $0.011\overline{011} \rightarrow 0.11\overline{011} \rightarrow 0.1\overline{011} \rightarrow 0.\overline{011}$
5. $0.1\overline{1} \rightarrow 0.\overline{1}$
6. $0.\overline{10} \rightarrow 0.0\overline{10} \rightarrow 0.\overline{10} \rightarrow 0.0\overline{10}$
7. $0.1110\overline{110} \rightarrow 0.110\overline{110} \rightarrow 0.10\overline{110} \rightarrow 0.\overline{110}$
8.

Third stage	0.000...	0.001...	0.010...	0.011...	0.100...	0.101...	0.110...	0.111...
	000	001	010	011	100	101	110	111

9. In 110 and 11011
10.–11. $x_1 = 0.10100\overline{110}$ in 101 and 1010, $x_2 = 0.0100\overline{110}$ in 010 and 0100
   $x_3 = 0.100\overline{110}$ in 100 and 1001, $x_4 = 0.00\overline{110}$ in 001 and 0011

ACTIVITY 5.5

1. $x_1 = 0.00\overline{110}$ in 001, $x_2 = 0.0\overline{110}$ in 011, $x_1 = 0.\overline{110}$ in 110
2. x_0 in 010, x_1 in 101, x_2 in 010, x_3 in 100
   x_4 in 001, x_5 in 011, x_6 in 111
3. $x_0 = 0.\overline{111010}$
   For example 0.01000111, 0.01001111, 0.01010111 and 0.01011111
4. The sixth iteration
5. For example 0.010101100 and 0.010111100

6.

	Stage-2 Interval	Stage-5 Interval	Stage-n Interval
x_0	$b_1 b_2$	$b_1 b_2 b_3 b_4 b_5$	$b_1 b_2 ... b_n$
$x_2 = S^2(x_0)$	$b_3 b_4$	$b_3 b_4 b_5 b_6 b_7$	$b_3 ... b_n c_1 c_2$
$x_5 = S^5(x_0)$	$b_6 b_7$	$b_6 b_7 b_8 b_9 b_{10}$	$b_6 ... b_n c_1 ... c_5$
$x_n = S^n(x_0)$	$c_1 c_2$	$c_1 c_2 c_3 c_4 c_5$	$c_1 c_2 ... c_n$

7.

000	001	010	011	100	101	110	111
x_{14}	x_1	x_5	x_2	x_0	x_4	x_3	

8. For example $x_0 = 0.100110101001010000111$ at 18
9. 10
10. $1/3 \rightarrow 2/3 \rightarrow 1/3$; period 2
11. $2/10 = 1/5 \rightarrow 2/5 \rightarrow 4/5 \rightarrow 3/5 \rightarrow 1/5$; period 4
12. $2/9 \rightarrow 4/9 \rightarrow 8/9 \rightarrow 7/9 \rightarrow 5/9 \rightarrow 1/9 \rightarrow 2/9$; period 6
13. $3/7 \rightarrow 6/7 \rightarrow 5/7 \rightarrow 3/7$; period 3
14. $3/15 = 1/5$; period 4
15. $26/31$; period 5
16. $x_2 = 0.\overline{a_3 a_4 a_1 a_2}$, $x_3 = 0.\overline{a_4 a_1 a_2 a_3}$, $x_4 = 0.\overline{a_1 a_2 a_3 a_4}$; period 4
17. $0.\overline{011}$, $0.\overline{001}$ and $0.\overline{110}$
18. $0.\overline{1101}$; subintervals 11011, 10111, 01110 and 11101
19. $0.\overline{000\ 001\ 010\ 011\ 100\ 101\ 110\ 111}$ is such a point
20. Delete all digits "1" if they occure in packets of 3 or more
21. $(1/2)^n < 1/50$ implies $n = 6$
22. $n = 8$

23. Smallest $x_s = 0.101_{two}$; largest $x_l = 0.101\overline{1}_{two} = 0.110$; size 1/8, 10110 contains x_0, size 1/32
24. $x_0 - z_0 = 0.00001$, distance 1/32
25. $|S(x_0) - S(z_0)| = 0.0001$, distance 1/16
26. $|S^2(x_0) - S^2(z_0)| = 0.001$, distance 1/8
27. $0.1\overline{110}$
28. $0.\overline{001}$
29. $|S^3(x_0) - S^3(z_0)| = 1/2$
30. $x_0 = 0.10110$, $z_0 = 0.10111$

ACTIVITY 5.6

1. $2.654\overline{26}$
2. $9.6\overline{6}$
3. $9.\overline{876543}$
4. $0.101\overline{10}$
5. $1.010\overline{011}$
6. $0.0100\overline{1}$
7. $0.110\overline{110} \rightarrow 0.01\overline{001}$
8. $3/11 \rightarrow 6/11$
9. $0.35_{tem} \rightarrow 0.70_{ten}$
10. $0.10\overline{100} \rightarrow 0.1\overline{011} \rightarrow 0.\overline{100} \rightarrow 0.10\overline{100}$
11. $0.11\overline{01} \rightarrow 0.0\overline{10} \rightarrow 0.\overline{10} \rightarrow 0.\overline{10}$
12. $0.001101 \rightarrow 0.01101 \rightarrow 0.1101 \rightarrow 0.010$

13.
$x = 0.10101$ $\qquad$ $x = 0.10101$
$T(x) = 0.1010\overline{1}$ $\qquad$ $S(x) = 0.0101$
$T(T(x)) = 0.101$ $\qquad$ $T(S(x)) = 0.101$

$x = 0.01101$ $\qquad$ $x = 0.01101$
$T(x) = 0.1101$ $\qquad$ $S(x) = 0.1101$
$T(T(x)) = 0.010\overline{1}$ $\qquad$ $T(S(x)) = 0.010\overline{1}$

14.

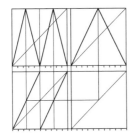

15. They are the same

16.

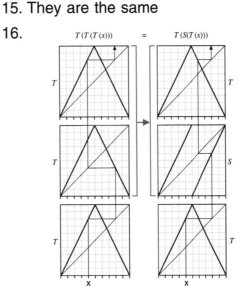

20. $T^5(x_0) = T(S^4(x_0)),\ S^4(x_0) = 0.\overline{01100},\ T(S^4(x_0)) = 0.\overline{11000}$
    $0.110\overline{01100} \rightarrow 0.011\overline{10011} \rightarrow 0.11\overline{10011} \rightarrow 0.00\overline{1100} \rightarrow 0.0\overline{1100} \rightarrow 0.\overline{11000}$

ACTIVITY 5.7

1. 011
2. $S^2(x_0) = 0.1101$ in 1101
3. right half

4. $T(S^2(x_0)) = 0.010\overline{1}$ in 010
5. $c_1^* c_2^* c_3^*$

6. 110

7. $S^2(x_0) = 0.0010$ in 0010
8. left half

9. $T(S^2(x_0)) = 0.010$
10. $c_1 c_2 c_3$
11. $S^{n-1}(x_0) = 0.1 b_1^* b_2^* ... b_n^*,\ T(S^{n-1}(x_0)) = 0.b_1 b_2...b_n \overline{1}$ (endpoint of interval $b_1 b_2...b_n$)
12. $S^{n-1}(x_0) = 0.0 b_1 b_2...b_n,\ T(S^{n-1}(x_0)) = 0.b_1 b_2...b_n$ (start of interval $b_1 b_2...b_n$)
13. $x_0 = .10011001$
14. 3
15. 4
16. 2

Iteration by T

17. $0.\overline{0100011}$
18. $0.\overline{11010}$
19. $0.1 = 1/2$
20. $x_n = 0.a_{n+1}^* a_{n+2}^* a_{n+3}^* \cdots,\ \ z_n = 0.a_{n+1}^* a_{n+2}^* a_{n+3}^* \cdots$; distance 1/2
21. $x_n = 0.101101,\ \ z_n = 0.101111$

ACTIVITY 5.8

1.

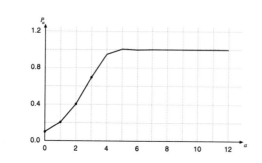

2.

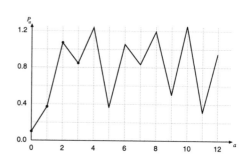

3. The first settles down at 1.0. The second oscillates in an unpredictable way.
4. Increase if $P < 1$, decrease if $P > 1$.
7.   8. Predictable oscillation

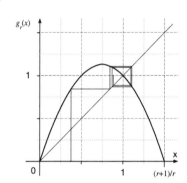

ACTIVITY 5.9

2. At 3/2
3. Allways at 1
4. $3/8 = 0.375 \to 0.844 \to 1.107 \to 0.869 \to 1.096$
6. $1/4 = 0.25 \to 0.563 \to 0.738 \to 0.580 \to 0.731$
5.   7.

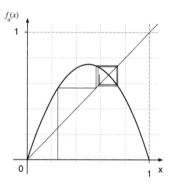

8. $3/8 = 0.375 \to 0.845 \to 1.107 \to 0.870 \to 1.097$
9. $0.3 \to 0.930 \to 1.125 \to 0.702 \to 1.330$
10. $0.225 \to 0.698 \to 0.844 \to 0.527 \to 0.997$

11.

x_0	x_{10}	x_{20}	x_{30}	x_{40}	x_{50}
0.3	0.102	1.321	0.927	1.214	0.317

12.

x_0	x_{10}	x_{20}	x_{30}	x_{40}	x_{50}
0.225	0.077	0.991	0.695	0.980	0.967

ACTIVITY 5.10

1. $0.08 \to 0.294 \to 0.831 \to 0.562 \to 0.985 \to 0.061$
   $0.10 \to 0.360 \to 0.922 \to 0.289 \to 0.822 \to 0.585$
2. Sensitivity

3.–7.

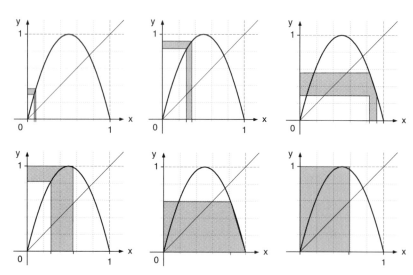

8. Mixing
9. p_0: period 1, q_0: period 2, r_0: period 3, s_0: period 10
10. $x_0 = 0$, $x_0 = 3/4$
11. A fixed point satisfies $x_0 = f(x_0)$. Using this equation to substitute the argument of f in this same equation provides $x_0 = f(f(x_0))$.
12. First $x_0 = 0$ and $-64x_0^3 + 128x_0^2 - 80x_0 + 15 = 0$,
    then $(x_0 - 3/4)(-64x_0^2 + 80x_0 - 20) = 0$
13. $x_0 = f(f(f(x_0))) = f^3(x_0)$, $x_0 = f^4(x_0)$, $x_0 = f^n(x_0)$

ACTIVITY 5.11

1.

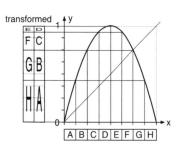

2. Smallest: D and E; largest: A and H

3.

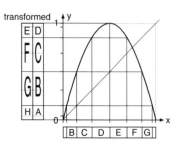

4. All cover 2 subdivisions

5. 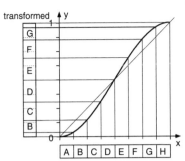   This is the subdivision of the x-axis of question 3.

6.–7. 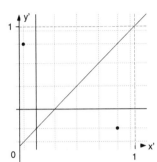   8.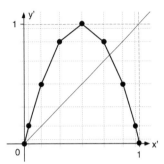

9.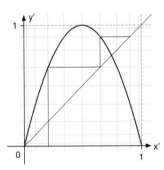

10. $x'_0 = 0.345$

11. $x'_0 = 0.18825$, $x'_1 = 0.61126$, $x'_2 = 0.95048$

12. x'_k : 0.232 0.713 0.819 0.594 0.965 0.136 0.469 0.996 0.016 0.062 0.032

    $f^k(x'_0)$: 0.232 0.713 0.819 0.594 0.965 0.136 0.469 0.996 0.016 0.062 0.032

13.a. $y_1 = 4\sin^2(x_0\pi/2)(1 - \sin^2(x_0\pi/2))$

13.b. $y_1 = 4\sin^2(x_0\pi/2)\cos^2(x_0\pi/2)$

13.c. $y_1 = \sin^2(x_0\pi)$

13.d. $\sin^2(x_1\pi/2) = \sin^2(x_0\pi)$

13.e. $\sin^2(x_1\pi/2) = \sin^2(\pi - x_0\pi)$

13.f. $\sin^2(x_1\pi/2) = \sin^2(-x_0\pi) = \sin^2(x_0\pi)$

ACTIVITY 5.12

1. Left: path leads to fixed point;
   right: path spirals around in an unpredictable way; no
2. Left: stable; right: unpredictable

3. $a = 1.9$: settles down on fixed point
4. $a = 2.75$: oscillates, but settles down on fixed point
5. $a = 3.25$: settles down on 2-cycle
6. Change from fixed point to 2-cycle; no
7. $a = 3.50$: settles down on 4-cycle
8. $a = 3.75$: shows unpredictable oscillation
9. $a = 4.00$: shows unpredictable oscillation over the whole unit interval
10. Change from predictable (found up to $a = 3.5$) to unpredictable

ACTIVITY 5.13

1.
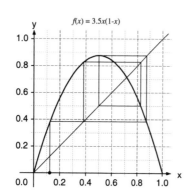

3.
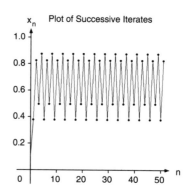

4. $a = 3.18$: settle on 2-cycle
5. $a = 3.5$: settle on 4-cycle
6. $a = 4.$: wild oscillation over the whole unit interval
7. $a = 2.8$: fixedpoint
8. $a = 3.18$: 2-cycle
9. $a = 3.5$: 4-cycle
10. $a = 4.0$: fills complete unit interval $[0, 1]$
11.
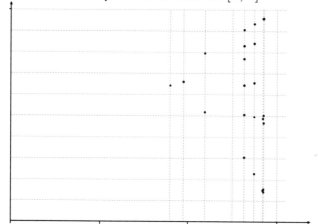

12.

Parameter a:	2.95	3.63	3.74	3.84	3.846
Points that form the attractor	0.661	0.305	0.935	0.959	0.961
		0.770	0.227	0.149	0.142
		0.644	0.657	0.488	0.470
		0.832	0.843		0.958
		0.506	0.496		0.155
		0.907			0.503

13. Fixed point attractors for $a \leq 3$; 2-cycles for parameters in neighborhood of $a = 3.18$;

4-cycles for parameters in neighborhood of $a = 3.5$; 3-cycles for parameters in neighborhood of $a = 3.84$;

14. The long time behavior changes more rapidly.

15.
$$
\begin{aligned}
d_1 &= b_2 - b_1 = 3.449489... - 3.0 &&\approx 4.4949 \cdot 10^{-1} \\
d_2 &= b_3 - b_2 = 3.544090... - 3.449490... &&\approx 9.4611 \cdot 10^{-2} \\
d_3 &= b_4 - b_3 = 3.564407... - 3.544090... &&\approx 2.0316 \cdot 10^{-2} \\
d_4 &= b_5 - b_4 = 3.568759... - 3.564407... &&\approx 4.3521 \cdot 10^{-3} \\
d_5 &= b_6 - b_5 = 3.569692... - 3.568759... &&\approx 9.3219 \cdot 10^{-4} \\
d_6 &= b_7 - b_6 = 3.569891... - 3.569692... &&\approx 1.9964 \cdot 10^{-4} \ .
\end{aligned}
$$

16.
$$
\begin{aligned}
d_1/d_2 &= 4.7514... \\
d_2/d_3 &= 4.6562... \\
d_3/d_4 &= 4.6682... \\
d_4/d_5 &= 4.6687... \\
d_5/d_6 &= 4.6690...
\end{aligned}
$$

UNIT 6   The Mandelbrot Set

ACTIVITY 6.1

1.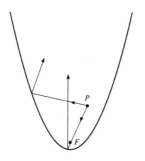

2.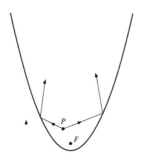

3. The marked focus.

4.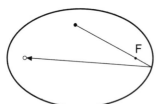

5.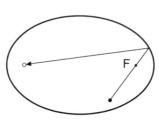

6. Either through the focus F or directly to the pocket.

7. The focus F.

8.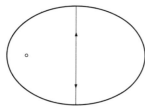

9.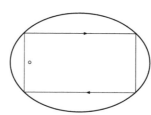

10. 2 and 4, respectively.

11. Yes.

12. b and c.

13. a. $x = 0$, b. $y = \frac{2}{5}x + \frac{6}{5}$, c. Any linear equation satisfied by (0,3).

14. (-5,0)

15. a and b.

16. a. $x = 2$ and $y = \frac{3}{2}x + 1$, b. $x = -3$ and $y = -\frac{5}{3}x + 1$, c. $x = -\frac{1}{2}$ and $y = -x + 1$

ACTIVITY 6.2

1.

x_0	$x_1 = f(x_0)$	$x_2 = f(x_1)$	$x_3 = f(x_2)$	$x_4 = f(x_3)$	Prisoner
−0.5	−1	−2	−4	−8	No
2.75	1.5	2.5	2	3	
3.0	1	2	3	1	
4.5	5.5	8.5	17.5	44.5	No

3. 0.75

4. 3

5. 3
6. Range interval [0,3] is not a subset of domain interval [0,2].
7. Range interval [0,3] is a subset of domain interval [0,3].
8. Range interval [1,4] is a subset of domain interval [1,4].
9.

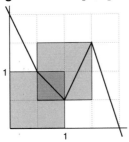

10. Case b.
11. [0,4] for both graphs.

ACTIVITY 6.3

1.

Stage 0	
Stage 1	
Stage 2	
Stage 3	
Stage 4	

2. 32; 64; 2^n
3. $(1/3)^5$; $(1/3)^6$; $(1/3)^n$
4.

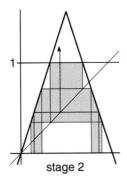

stage 2

5. $1/9 \to 1/3$; $2/9 \to 2/3$
6.–7.

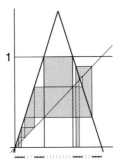

8. $25/9 \rightarrow 2/9 \rightarrow 2/3$ and $26/27 \rightarrow 1/9 \rightarrow 1/3$.
9. 7; 15
10. $E = 2^N - 1$. As N increases towards infinity, E also increases.

ACTIVITY 6.4

1. For all $x \in [0, 1]$, $f(x) \in [0, 1]$.
2. For all $x \in [0, 1]$, $f(x) \in [0, 1]$.
3. For $x = 0.5$, $f(x) \notin [0, 1]$.

4.

Function $f(x)$	x_0	$x_1 = f(x_0)$	$x_2 = f(x_1)$	$x_3 = f(x_2)$
$f(x) = 2x(1 - x)$	-0.5	-1.5	-7.5	-127.5
	1.5	-1.5	-7.5	-127.5
$f(x) = 4x(1 - x)$	-0.5	-3	-48	-9408
	1.5	-3	-48	-9408
$f(x) = 6x(1 - x)$	-0.5	-4.5	-148.5	-133204.5
	1.5	-4.5	-148.5	-133204.5
	0.9	0.54	1.4904	-4.3854

5. In each case, the sequence iterates towards negative infinity.
6. The maximal invariant interval for $f(x) = 2x(1 - x)$ is [0,1].
    The maximal invariant interval for $f(x) = 4x(1 - x)$ is [0,1].
    No invariant interval exists for $f(x) = 6x(1 - x)$.
7. Connected.
8.–9.

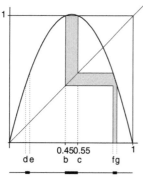

10. To solve $f(d) = b$ with $b = 0.45$ compute
    $$\frac{400}{99} d(1 - d) = 0.45$$
    $$d - d^2 = 0.111375$$
    $$d^2 - d + 0.111375 = 0$$
    $$d = 0.5 \pm \sqrt{0.25 - 0.111375}$$
    which yields $d \approx 0.1277$. The same procedure produces $e \approx 0.1625$, $f \approx 0.8375$, and $g \approx 0.8723$.

11.–13.

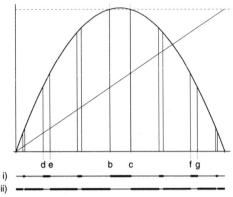

i)
ii)

14. a. 15, b. 16.
15. 0.

ACTIVITY 6.5

1.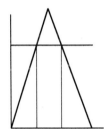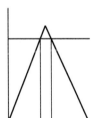

2. The width of the escape interval decreases.
3. Maximum height $f(0.5) = 3$, $(b,c) = (1/6, 5/6)$, escape interval width 2/3.
4. $f(0.5) = 1$.
5. $a = 2$.
6.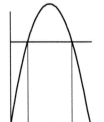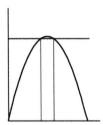

7. Maximum height $f(0.5) = 1$.
8. $a = 4$.
9. When points from within [0,1] are iterated through functions in which the parameter a is less than the barrier, the iterates cannot escape the interval [0,1]. When the parameter exceeds the barrier, whole intervals exist which contain points that excape from [0,1] under iteration.
10. The point $x = 0.5$ is associated with the maximum height of the function f. The height $f(0.5)$ determines whether or not points can escape. The maximum point for $f(x) = ax(1-x)$ also occurs at $x = 0.5$. Accordingly, $x = 0.5$ also acts as the critical initial point for this function.

11. $f(x) = ax(1 - x^2) = ax^3 - ax$ for $x \geq 0$. Compute $f'(x) = 3ax^2 - a$ and solve $3ax^2 - a = 0$ which yields $x^2 = 1/3$ and $x = \sqrt{3}/3$, the critical point, and $(\sqrt{3}/3, 2a\sqrt{3}/9)$, the relative maximum.

ACTIVITY 6.6

1. **a.** $f(x) = x^2 - 0.25$ **b.** $f(x) = x^2 - 0.75$ **c.** $f(x) = x^2 - 2$

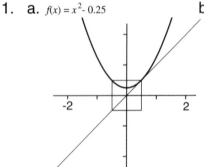 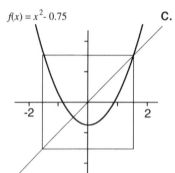 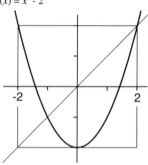

2. The parabola extends below the box for initial points near $x = 0$. The box test implies that these points escape and no Invariant interval exists. In the absence of an Invariant interval for such functions as these, a Cantor Set of prisoner points results.

3. The negative y-intercept implies that $c < 0$. At the intersection of the parabola and the diagonal we have $x^2 + c = x$, thus $x^2 - x + c = 0$. By the quadratic formula, $x = 1 \pm \sqrt{1 - 4c}$. Since $c < 0$, only $x = 1 + \sqrt{1 - 4c}$ is in the first quadrant.

4. $f(x) = x^2 + c = x$ and $x = -c$ yields $c^2 + c = -c$ or $c(c + 2) = 0$ which has only one positive solution, $c = -2$.

5. $c = -2$ is the barrier.

ACTIVITY 6.7
1.

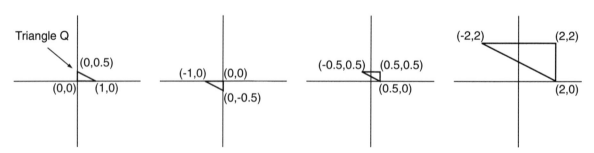

2. Triangle P $(1,0) \to (-1.5, 1.5)$ $(1,2) \to (-1.5, -4.5)$
    Triangle Q $(2,0) \to (-6, 2)$ $(0,1) \to (2, -2)$.

3. $(0.5, 0.75) \rightarrow (0, -0.75)$
    $(0,0) \rightarrow (1.5, 1.5)$
    $(1,0) \rightarrow (-1.5, 1.5)$

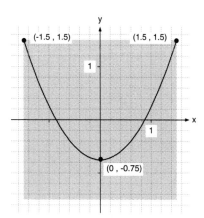

4. a.

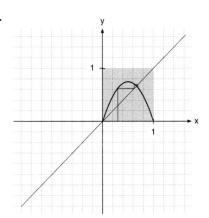

    b.

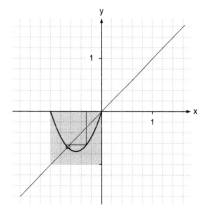

4. c.

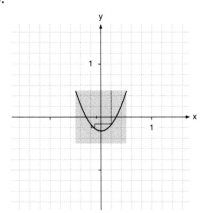

    d.

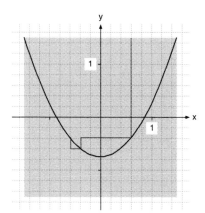

5. The two graphs are identical.

6.

Original	$y = 1x(1-x)$	$y = 3x(1-x)$	$y = 4x(1-x)$	$y = ax(1-x)$
Image	$y = x^2 + \frac{1}{4}$	$y = x^2 - \frac{3}{4}$	$y = x^2 - 2$	$y = x^2 + \left(\frac{a}{2} - \frac{a^2}{4}\right)$

7. See answer 4

9.

Original	$y = 1x(1-x)$	$y = 3x(1-x)$	$y = 4x(1-x)$	$y = ax(1-x)$
Prisoner set	$[0, 1]$	$[0, 1]$	$[0, 1]$	$[0, 1]$
Image	$[-1/2, 1/2]$	$[-3/2, 3/2]$	$[-2, 2]$	$[-a/2, a/2]$

ACTIVITY 6.8

1. [Entering code in appropriate graphics calculator.]

2.

Value of c	−3	−2.1	−2.01	−2.001	−2.0001	−2.00001	−2.000001
Iterations N	4	6	8	10	11	13	15

3. The number of iterations N is proportional to the logarithm of $1/(2-c)$.

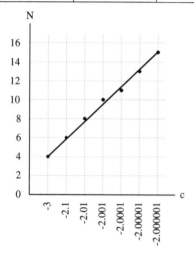

4. As the value of parameter c increases towards −2, the number of iterates required before their value exceeds 1000 also increases.

5.

−3	−2.1	−2.01	−2.001	−2.0001	−2.00001	−2.000001
yellow	orange	brown	red	red	green	blue

ACTIVITY 6.9

1. a. $3+3i$, b. $2+i$, c. $4-3i$, d. $-2+2i$.

2.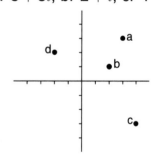

3.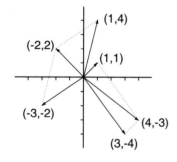

4. a. $(1+2x)(3+5x)$ $3+11x+10x^2$ $3+11i+10i^2$ $-7+11i$
    b. $(3-4x)(5-7x)$ $15-41x+28x^2$ $15-41i+28i^2$ $-13-41i$
    c. $(a+bx)(c+dx)$ $ac+(ad+bc)x+bdx^2$ $ac+(ad+bc)i+bdi^2$ $(ac-bd)$
    $+(ad+bc)i$

5. $|v| = \sqrt{2^2+1^2} = \sqrt{5} \approx 2.236.$ $|w| = \sqrt{3^2+4^2} = \sqrt{25} = 5.$

6.

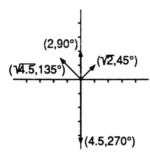

7. $(1+i)(1+i) = (1-1) + (1+1)i = 2i.$
$(3/2 + (3/2)i)(3/2 + (3/2)i) = (9/4 - 9/4) + (9/4 + 9/4)i = (9/2)i.$

8. **a.** $(1/2+i)(2+i) = (5/2)i$
    b. $(1+i)(0+2i) = -2+2i$

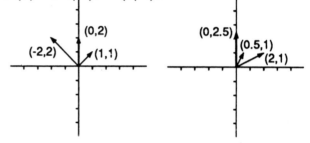

9.

	$1/2+i$	$2+i$	$0+5/2i$	$1+i$	$0+2i$	$-2+2i$
angle	60	30	90	45	90	135
size	$\sqrt{5}/2$	$\sqrt{5}$	5/2	$\sqrt{2}$	2	$\sqrt{8}$

10. The product of two vectors has a length equal to the product of the two lengths and an angle equal to the sum of the two angles.

ACTIVITY 6.10

1. 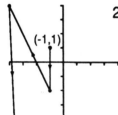   **2.** 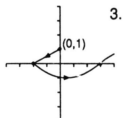   **3.** 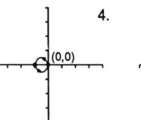   **4.**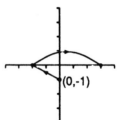

5. [Entering code in appropriate graphing calculator.]

6. 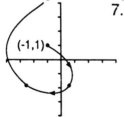   **7.** 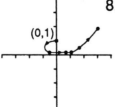   **8.** 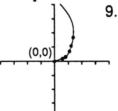   **9.**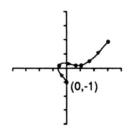

10. The prisoner set is totally disconnected consisting of a relatively scarce "dust" of points.

ACTIVITY 6.11

1. $0 + 0i \rightarrow 0 + i \rightarrow -1 + i \rightarrow 0 - i \rightarrow -1 + i \rightarrow 0 - i \rightarrow \cdots$
   After two steps the sequence of iterates oscillates forever between two points.
2. $0 + 0i \rightarrow 0.5 + 0.1i \rightarrow 0.74 + 0.2i \rightarrow 1.0076 + 0.396i \rightarrow 1.358 + 0.898i \rightarrow$
   $1.539 + 2.5398i \rightarrow -3.582 + 7.917i \rightarrow -49.3 - 56.6i \rightarrow \cdots$
   The sequence of iterates continues to increase in absolute magnitude.
3. Using $f(z) = z2 + (-1 + 0i)$, the sequence of iterates associated with the critical point $0 + 0i$ oscillates between two values, $0 + 0i$ and $-1 + 0i$. Hence, $0 + 0i$ fails to escape and the prisoner set is connected.
4. $0 + 0i \rightarrow -1 - i \rightarrow -1 + i \rightarrow -1 - 3i \rightarrow -9 + 5i \rightarrow 55 - 91i \rightarrow \cdots$ Using $f(z) = z^2 + (-1 - 1i)$, the sequence of iterates associated with $0 + 0i$ escapes towards infinity. Hence, the prisoner set is a totally disconnected Cantor set.
5. a. Connected.
   b. The sequence of points lies ever closer to the origin.
   c. The sequence of points moves out towards infinity.
   d. The Julia Set is the circle centered at the origin with radius 1.
6. a. Prisoner.
   b. Prisoner.
   c. Prisoner.
   d. Not a prisoner.
7. If $f(z) = z^2 + (-1 + 0i)$, then $z = (1 \pm \sqrt{5})/2 + 0i$.

ACTIVITY 6.12

1.

Parameter c	
$0.2 + 0.2i$	P
$-0.3 + 0.2i$	P
$-0.6 + 0.4i$	P
$-1.3 + 0.1i$	E
$-1.7 + 0.1i$	E

Parameter c	
$0.2 - 0.2i$	P
$-0.3 - 0.2i$	P
$-0.6 - 0.4i$	P
$-1.3 - 0.1i$	E
$-1.7 - 0.1i$	E

2.–5.

ACTIVITY 6.13

1.

Parameter c	
$0.2 + 0.2i$	P
$-0.3 + 0.2i$	P
$-0.6 + 0.4i$	P
$-1.3 + 0.1i$	17
$-1.7 + 0.1i$	8

Parameter c	
$0.2 - 0.2i$	P
$-0.3 - 0.2i$	P
$-0.6 - 0.4i$	P
$-1.3 - 0.1i$	17
$-1.7 - 0.1i$	8

2.–4.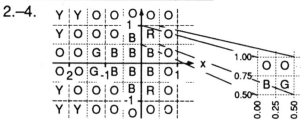

5. For real valued parameters c within $[-\infty, -2)$, the prisoner sets will be totally disconnected. When parameter c is within $[-2, 1/4]$, the prisoner set is connected. For real parameters $c > 1/4$ the prisoner set is disconnected and has no points on the real axis.